An Introduction
to Lebesgue Integration
and Fourier Series

HOWARD J. WILCOX

Professor of Mathematics
Wellesley College

and

DAVID L. MYERS

Mathematics Teacher
The Winsor School

DOVER PUBLICATIONS, INC.
New York

31143007178941
515.43 Wil
Wilcox, Howard J.
An introduction to
Lebesgue integration and
Fourier series
Dover ed.

Bibliographical Note

This Dover edition, first published in 1994, is an unabridged, slightly
corrected republication of the work first published by the Robert E. Krieger
Publishing Company, Huntington, New York, 1978 (in the "Applied Mathematics Series").

Library of Congress Cataloging in Publication Data

Wilcox, Howard J.
    An introduction to Lebesgue integration and Fourier series / Howard J.
Wilcox and David L. Myers.
        p.   cm.
    Originally published: Huntington, N.Y. : R. E. Krieger Pub. Co., 1978, in
series: Applied mathematics series.
    Includes bibliographical references and index.
    ISBN 0-486-68293-5 (pbk.)
    1. Integrals, Generalized.   2. Fourier series.   I. Myers, David L.
II. Title.
QA312.W52   1994
515'.43—dc20   94-35540
                                                                  CIP

Manufactured in the United States of America
Dover Publications, Inc., 31 East 2nd Street, Mineola, N.Y. 11501

Contents

Preface

This book arose out of our desire to present an introduction to Lebesgue Integration and Fourier Series in the second semester of our real variables course at Wellesley College. We found that most undergraduate texts do not cover these topics, or do so only in a cursory way. Graduate texts, we felt, lack motivation and depend on a level of sophistication not attained by most undergraduates.

We feel this text could be used for a course lasting one semester or less (there are several optional sections, marked with an asterix, which could easily be omitted). We assume knowledge of advanced calculus, including the notions of compactness, continuity, uniform convergence and Riemann integration (i.e., a usual one-semester undergraduate course in advanced calculus). Therefore, the book would be suitable for advanced undergraduates and beginning graduate students.

It is our intention throughout the book to motivate what we are doing. Goals of the theory are kept before the reader, and each step of the development is justified by reference to them. For example, the inadequacies of the Riemann integral are pointed out, and each new accomplishment of the Lebesgue theory is measured against the goal of overcoming these difficulties. In addition, each new concept is related to concepts already in the student's repertoire, whenever this is possible. The Lebesgue integral is defined in terms precisely analogous to the Riemann-sum definition of the Riemann integral. The primary difference is that in Lebesgue's approach, "Lebesgue sums" are formed relative to an arbitrary partition of an interval containing the range of a bounded function, in contrast to Riemann's partitioning of the domain.

The formation of "Lebesgue sums" leads naturally to the goal of defining the measure of an arbitrary set. Outer measure is defined in the classical way, and is shown to lack countable additivity on the collection of all subsets of $[0,1]$. This leads to the restriction of attention to measurable sets, and hence to measurable functions. The theory is pursued through the usual convergence results, which overcome some of the deficiencies of the Riemann theory. This is followed by a discussion of linear spaces and $\mathcal{L}^2$ in particular. This leads in a natural way to the $\mathcal{L}^2$ theory of Fourier series. Finally, pointwise convergence of Fourier series is discussed.

CHAPTER *1*

The Riemann Integral

1. Definition of the Riemann Integral

The problem of finding the area of a plane region bounded by vertical lines $x = a$ and $x = b$, the horizontal line $y = 0$, and the graph of the non-negative function $y = f(x)$, is a very old one (although, of course, it has not always been stated in this terminology). The Greeks had a method which they applied successfully to simple cases such as $f(x) = x^2$. This "method of exhaustion" consisted essentially in approximating the area by figures whose areas were known already—such as rectangles and triangles. Then an appropriate limit was taken to obtain the result.

In the seventeenth century, Newton and Leibnitz independently found an easy method for solving the problem. The area is given by $F(b) - F(a)$, where F is an antiderivative of f. This is the familiar Fundamental Theorem of Calculus; it reduced the problem of finding areas to that of finding antiderivatives.

Eventually mathematicians began to worry about functions not having antiderivatives. When that happened, they were forced to return again to the basic problem of area. At the same time, it became clear that a more precise formulation of the problem was necessary. Exactly what is area, anyway? Or, more generally, how can $\int_a^b f(x)dx$ be defined rigorously for as wide a class of functions as possible?

In the middle of the nineteenth century, Cauchy and Riemann put the theory of integration on a firm footing. They described—at least theoretically—how to carry out the program of the Greeks for any function f. The result is the definition of what is now called the *Riemann integral* of f. This is the integral studied in standard calculus courses.

1.1 Definition: A *partition* P of a closed interval $[a,b]$ is a finite sequence $(x_0, x_1, \ldots, x_n)$ such that $a = x_0 < x_1 < \ldots < x_n = b$. The *norm* (or *width*, or *mesh*) of P, denoted $\|P\|$, is defined by

$$\|P\| = \max_{1 \leqslant i \leqslant n} (x_i - x_{i-1}).$$

That is, $\|P\|$ is the length of the longest of the subintervals $[x_0, x_1]$, $[x_2, x_3], \ldots, [x_{n-1}, x_n]$.

1.2 Definition: Let $P = (x_0, \ldots, x_n)$ be a partition of $[a,b]$, and let f be defined on $[a,b]$. For each $i = 1, \ldots, n$, let x_i^* be an arbitrary point in the interval $[x_{i-1}, x_i]$. Then any sum of the form

$$R(f,P) = \sum_{i=1}^{n} f(x_i^*)(x_i - x_{i-1})$$

is called a *Riemann sum of f relative to P.*

Notice that $R(f,P)$ is not completely determined by f and P; it depends also on the choice of the elements x_i^*. For a non-negative function f, $R(f,P)$ is the sum of the areas of rectangles approximating the area under the graph of f (see diagram).

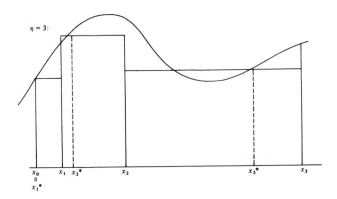

Now we take a limit of our approximating areas.

1.3 **Definition:** A function f is *Riemann integrable* on $[a,b]$ if there is a real number R such that for any $\epsilon > 0$, there exists a $\delta > 0$ such that for any partition P of $[a,b]$ satisfying $\|P\| < \delta$, and for any Riemann sum $R(f,P)$ of f relative to P, we have $|R(f,P) - R| < \epsilon$.

We can rewrite this in logical shorthand as: f is integrable on $[a,b]$ if there is a number R such that

$$(\forall \epsilon > 0)(\exists \delta > 0)(\forall P)[\|P\| < \delta \to |R(f,P) - R| < \epsilon],$$

where we must remember the meanings of P and $R(f,P)$.

What is this number R? First of all, there can be at most one number R which satisfies the condition. For, suppose that R and R' both worked; then we could take $\epsilon = (1/2)|R - R'|$. As the reader can verify (Exercise 5.2), this would entail the existence of a Riemann sum $R(f,P)$ satisfying both $|R(f,P) - R| < \epsilon$ and $|R(f,P) - R'| < \epsilon$. An application of the triangle inequality yields a contradiction. Since at most one number R can satisfy our definition, and it is evidently the "limit" of the Riemann sums, we define it, if it exists, to be the Riemann integral of f on $[a,b]$, denoted $\int_a^b f(x)dx$. (For f non-negative, we also define this number to be the area we have described above.)

The fact that the definition of integrability is a kind of limit will be emphasized if we write down the usual definition of a function H having a limit as x approaches 0:

there is a real number b such that

$$(\forall \epsilon > 0)(\exists \delta > 0)(\forall x \neq 0)[|x| < \delta \to |H(x) - b| < \epsilon],$$

where we understand that x is restricted to the domain of H. Since we write $\lim_{x \to 0} H(x) = b$ in this case, we might write

$$\lim_{\|P\| \to 0} R(f,P) = \int_a^b f(x)dx$$

for the integral. But we must keep in mind that this is only an abbreviation of the (necessarily) involved $\epsilon - \delta$ definition. The difference between the two types of limits, arising mainly from the fact that $R(f,P)$ is not a function simply of $\|P\|$, whereas H *is* a function of x, should also be noted.

The definition of the Riemann integral given above has a drawback which it shares with all limit definitions—to prove that a particular function is integrable, you must first know the value of $R = \int_a^b f(x)dx$. The definition itself gives no direct means of finding R. If we restrict ourselves for the moment to non-negative functions, more information about the area R can be obtained by considering rectangles lying entirely below the graph of f, and rectangles whose tops lie above the graph of f. Then it would seem reasonable that the area R should lie between the area of the inner rectangles and the area of the outer rectangles. In fact, it turns out that f is integrable if and only if inner and outer rectangles can be

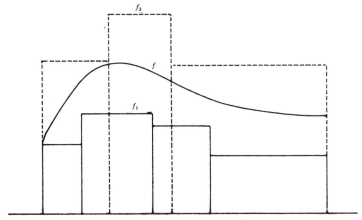

found whose total areas are arbitrarily close to each other. This is the content of Theorem 1.6 below. Note that this theorem holds for any function (not necessarily non-negative).

Before we state the theorem, we need to introduce some useful notation. Note that the tops of the outer rectangles define a function, except for some ambiguity at the shared boundaries between two rectangles (and similarly for the inner rectangles). Furthermore, such a function has a particularly nice form.

1.4 Definition: A function g, defined on $[a,b]$, is a *step function* if there is a partition $P = (x_0, x_1, \ldots, x_n)$ such that g is constant on each open sub-interval (x_{i-1}, x_i), for $i = 1, \ldots, n$. (The values of $g(x_0), \ldots, g(x_n)$ are irrelevant.)

Notice that a step function on $[a,b]$ has finitely many values. Furthermore, a step function is Riemann integrable, and the value of its integral is the obvious area given in the following proposition.

1.5 Proposition: Any step function g on $[a,b]$ is Riemann integrable. Furthermore, if $g(x) = c_i$ for $x \in (x_{i-1}, x_i)$, where $(x_0, \dots, x_n)$ is a partition of $[a,b]$, then

$$\int_a^b g(x)dx = \sum_{i=1}^n c_i(x_i - x_{i-1})$$

(Notice that the values of g at $x_0, \dots, x_n$ have no effect on the integral.)

Proof: See Exercise 5.5.

□

Now we are ready for the theorem.

1.6 Theorem: A function f, defined on $[a,b]$, is Riemann integrable on $[a,b]$ if and only if for every $\epsilon > 0$, there are step functions f_1 and f_2 such that

$$f_1(x) \leqslant f(x) \leqslant f_2(x) \qquad \text{for all } x \in [a,b], \qquad \text{and}$$

$$\int_a^b f_2(x)dx - \int_a^b f_1(x)dx < \epsilon.$$

Proof: See Exercise 5.11. Note that if f is Riemann integrable, then f is bounded (Exercise 5.9).

□

Since the condition in the theorem is equivalent to integrability, we will use it and the definition interchangeably. We also have the following expressions for $\int_a^b f(x)dx$.

1.7 Corollary: If f is Riemann integrable on $[a,b]$, then

$$\int_a^b f(x)dx = \mathrm{lub}\left\{ \int_a^b f_1(x)dx \,\Big|\, f_1 \text{ a step function and } f_1 \leqslant f \right\}$$

$$= \mathrm{glb}\left\{ \int_a^b f_2(x)dx \,\Big|\, f_2 \text{ a step function and } f \leqslant f_2 \right\}.$$

2. Properties of the Riemann Integral

We state here for future reference some of the fundamental properties of the Riemann integral. Proofs of these can be found in any standard text on advanced calculus or analysis.

2.1 Theorem (Linearity): If f and g are Riemann integrable on $[a,b]$, so are cf (for any real number c), and $f + g$. Furthermore,

$$\int_a^b cf(x)dx = c \int_c^b f(x)dx,$$

and

$$\int_a^b [f(x) + g(x)]\,dx = \int_a^b f(x)dx + \int_a^b g(x)dx.$$

2.2 Theorem (Additivity): If $a < c < b$, then f is integrable on $[a,b]$ if and only if f is integrable on both $[a,c]$ and $[c,b]$. Furthermore,

$$\int_a^b f(x)dx = \int_a^c f(x)dx + \int_c^b f(x)dx.$$

2.3 Theorem (Monotonicity): If $f(x) \leqslant g(x)$ for all $x \in [a,b]$, and if f and g are Riemann integrable on $[a,b]$, then

$$\int_a^b f(x)dx \leqslant \int_a^b g(x)dx.$$

2.4 Corollary: If there are real constants m and M such that $m \leqslant f(x) \leqslant M$ for all $x \in [a,b]$, then

$$m(b-a) \leqslant \int_a^b f(x)dx \leqslant M(b-a).$$

2.5 Theorem: Any function which is continuous on $[a,b]$ is Riemann ingrable on $[a,b]$.

2.6 Theorem: Any function which is monotone on $[a,b]$ is Riemann integrable on $[a,b]$.

Using Theorem 2.2 and Exercise 5.5, we can extend the last two existence theorems to bounded *piecewise* continuous or bounded *piecewise* monotone functions.

Finally, the Fundamental Theorem of Calculus.

2.7 Theorem: Let U be an open interval containing $[a,b]$. If f is continuous on U and F is an antiderivative of f on U, then

$$\int_a^b f(x)dx = F(b) - F(a).$$

3. Examples

We have stated in the last section that any piecewise continuous or piecewise monotone function is Riemann integrable. We will now present two important examples of functions which are bounded but are neither piecewise continuous nor piecewise monotone. One will turn out to be Riemann integrable, and the other will not.

3.1 Definition: With any set of real numbers A, we associate a function χ_A, called the *characteristic function of A*, defined by

$$\chi_A(x) = \begin{cases} 1 & \text{if } x \in A \\ 0 & \text{if } x \notin A. \end{cases}$$

3.2 Example: Let Q be the set of rational numbers. Then χ_Q is bounded, but is neither piecewise continuous nor piecewise monotone, and it is not Riemann integrable on $[0.1]$. In fact, if f_1 and f_2 are step functions such that $f_1 \leq \chi_Q \leq f_2$, then there is an irrational x in each subinterval on which f_1 is constant, so that for that x,

$$f_1(x) \leq \chi_Q(x) = 0.$$

Therefore, except at finitely many points (the points of the partition for f_1), we have $f_1(x) \leq 0$, so that $\int_0^1 f_1(x)dx \leq 0$ by Corollary 2.4 and Exercise 5.5.

On the other hand, there is a rational number x in each interval on which f_2 is constant, so that $f_2(x) \geq \chi_Q(x) = 1$. Hence $\int_0^1 f_2(x)dx \geq 1$, and $\int_0^1 f_2(x)dx - \int_0^1 f_1(x)dx \geqslant 1$ cannot be made arbitrarily small, as required for integrability, by Theorem 1.6.

It is one of the advantages of the Lebesgue theory, to be introduced soon, that χ_Q will be Legesgue integrable. We will find its Lebesgue integral on $[0,1]$ to equal 0.

3.3 Example: Now we present a function which is neither piecewise monotone nor piecewise continuous, but which surprisingly enough is Riemann integrable. Let us agree for the purposes of this example to write non-zero rational numbers only in the form p/q, where $q \neq 0$ and p and q are integers having no factors in common (that is, p/q is written in "lowest terms"). Then define

$$g(x) = \begin{cases} 1/q & \text{if } x = p/q \\ 1 & \text{if } x = 0 \\ 0 & \text{if } x \text{ is irrational.} \end{cases}$$

Then g is continuous at each irrational and discontinuous at each rational (Exercise 5.23). Thus g is not quite as discontinuous as χ_Q, which is discontinuous at every point (Exercise 5.22). As part of our study of the Lebesgue integral, we will see just how discontinuous a Riemann integrable function can be (see Theorem 29.2). It is not too difficult to prove at this point that g is Riemann integrable on $[0,1]$ (Exercise 5.24).

4. Drawbacks of the Riemann Integral

The Riemann integral is adequate for most practical applications. The functions we usually encounter are piecewise continuous and very often have nice antiderivatives as well. However, in advanced theoretical investigations of functions of a real variable, we would like to have an integral with certain properties that the Riemann integral lacks.

First of all, there are certain easily described functions such as χ_Q which have no Riemann integral at all. We will greatly expand the range of integrable functions when we define the Lebesgue integral.

In addition, the Riemann integral lacks certain desirable limit properties. In particular, the class of Riemann integrable functions is not closed under pointwise limits, even for bounded monotone sequences.

4.1 Example: Let $Q \cap [0,1]$ be enumerated without duplications by $q_1, q_2, q_3, \ldots$. This is possible of course because Q is countable. Then define, for $n = 1, 2, \ldots$, and $x \in [0,1]$,

$$f_n(x) = \begin{cases} 1 & \text{if } x = q_1 \text{ or } x = q_2 \text{ or } \ldots \text{ or } x = q_n \\ 0 & \text{otherwise.} \end{cases}$$

The sequence $\{f_n\}$ is monotone increasing and is uniformly bounded (in fact $0 \leqslant f_n(x) \leqslant 1$ for all n and x). Each f_n is Riemann integrable, since f_n has exactly n discontinuities, hence is piecewise continuous. Finally, $\{f_n\}$ converges pointwise to χ_Q, a non-Riemann integrable function.

The Lebesgue integral not only overcomes many of these difficulties inherent in the Riemann integral, but—like any great mathematical idea—its study has also generated concepts and techniques which are extremely valuable to amthematicians doing research in diverse areas.

5. Exercises

5.1 Given $f(x) = x^3$ on $[0,1]$ and the partition $P = (0, 1/8, 1/3, 2/3, 1)$, find four different Riemann sums $R(f, P)$.

5.2 Prove that the number R in Definition 1.3 must be unique, if it exists.

5.3 Let $f(x) = c$ for all $x \in [a,b]$ and some real number c. Show by Definition 1.3 that f is Riemann integrable on $[a,b]$, and $\int_a^b f(x)dx = c(b - a)$.

5.4 Let $g(x) = 0$ for $x \neq 1$, $g(1) = 1$. Show from Definition 1.3 that $\int_0^2 g(x)dx = 0$. (Hint: given a partition $P = (x_0, x_1, \ldots, x_n)$ of $[0,2]$, $x = 1$ is in at most two subintervals $[x_{i-1}, x_i]$ and $[x_i, x_{i+1}]$. Thus show $R(g, P) \leqslant 2\delta$.)

5.5 (a) Prove that if f is Riemann integrable on $[a,b]$, $c \in [a,b]$, and $g(x) = f(x)$ for all $x \neq c$, then g is Riemann integrable on $[a,b]$, and $\int_a^b f(x)dx = \int_a^b g(x)dx$. (Hint: See Exercise 5.4.)
(b) Repeat (a) in the case where $g(x) = f(x)$ except at finitely many points $c_1, \ldots, c_k$ in $[a,b]$.

5.6 Let $f(x) = 0$ for $x \neq 1/n$, $n = 1, 2, 3, \ldots$, and let $f(1/n) = 1$. Show that $\int_0^1 f(x)dx = 0$.

5.7 Let $\sigma(x) = \begin{cases} 1 & \text{if } x > 0 \\ 0 & \text{if } x = 0 \\ -1 & \text{if } x < 0. \end{cases}$

(a)   Show that $f \circ \sigma$ is a step function for any function f defined on any closed interval $[a,b]$.

(b)   In contrast, find a function g such that $\sigma \circ g$ is not a step function on $[0,1]$.

(c)   Show that $\sigma \circ \sin$ is a step function on every closed interval, and find $\int_{-\pi/8}^{13\pi/8} \sigma(\sin x)dx$.

5.8   If f and g are step functions on $[a,b]$, show that $f + g$ is a step function. Is fg a step function? If $g:[a,b] \to [a,b]$, is $f \circ g$ a step function?

5.9   Prove that if f is Riemann integrable on $[a,b]$, then f is bounded, using only Definition 1.3. (The result follows easily from Theorem 1.6, but is needed in its proof.) (Hint: Assume that f is unbounded. Then relative to any partition P, $R(f,P)$ can be made arbitrarily large by making appropriate choices of $x_i^* \in [x_{i-1},x_i]$.)

5.10   Show that, in contrast to Exercises 5.5 and 5.6, there is a Riemann integrable function f on $[0,1]$ and a non-Riemann integrable function g on $[0,1]$ such that $f(x) = g(x)$ for all but countably many points in $[0,1]$. (Hint: see Exercise 5.9.)

5.11   Prove Theorem 1.6. (Hint: let f be Riemann integrable on $[a,b]$. Let ϵ, δ, P be as in Definition 1.3. Now if $f_1(x) = \text{glb}\{f(x) \mid x \in (x_{i-1},x_i)\}$ for $x \in (x_{i-1},x_i)$, then f_1 (defined appropriately at $x_0, \ldots, x_n$) is a step function lying below f. Show that there is a point $x_i^* \in [x_{i-1},x_i]$ such that $f(x_i^*) - f_1(x) < \epsilon$. Find f_2 above f in a similar manner, and $x_i^{**} \in [x_{i-1},x_i]$ with $f_2(x) - f(x_i^{**}) < \epsilon$.

For the converse, let $R = \text{lub}\{\int_a^b f_1(x)dx \mid f_1$ a step function and $f_1 \leqslant f\}$. Show that $R = \text{glb}\{\int_a^b f_2(x)dx \mid f_2$ a step function and $f \leqslant f_2\}$. Then given $\epsilon > 0$, obtain step functions $f_1 \leqslant f \leqslant f_2$ with $R - \int_a^b f_1(x)dx < \epsilon$ and $\int_a^b f_2(x)dx - R < \epsilon$. (Use the fact that f_1 and f_2 are integrable to obtain δ.)

5.12   The following is a popular alternative development of the Riemann integral. If f is a *bounded* function defined on $[a,b]$, and $P = (x_0, \ldots, x_n)$ is a partition of $[a,b]$, let $m_i = \text{glb}\{f(x) \mid x \in [x_{i-1},x_i]\}$, $M_i = \text{lub}\{f(x) \mid x \in [x_{i-1},x_i]\}$. Then $L(f,P) = \sum_{i=1}^n m_i(x_i - x_{i-1})$, $U(f,P) = \sum_{i=1}^n M_i(x_i - x_{i-1})$.

Let $\mathcal{L} = \{L(f,P) \mid P$ is a partition of $[a,b]\}$,

$U = \{U(f,P) \mid P$ is a partition of $[a,b]\}$.

(a)   Show that $\mathcal{L}$ is bounded above and U is bounded below.

Define $\int_a^b f(x)dx = \text{lub}\mathcal{L}$, and $\int_a^b f(x)dx = \text{glb}U$, called the *lower* and *upper* integrals of f, respectively.

(b)   Show that $\int_a^b f(x)dx \leqslant \int_a^b f(x)dx$.

(c)   Prove that f is Riemann integrable (in the sense of Definition 1.3 and Theorem 1.6) if and only if $\int_a^b f(x)dx = \int_a^b f(x)dx$. (In the alternative development using upper and lower sums, f is said to be integrable if the upper and lower integrals are equal.)

5.13 Prove Theorem 2.1. (Hint: use 1.6 and 1.7.)

5.14 Prove Theorem 2.2.

5.15 Prove Theorem 2.3.

5.16 Prove Corollary 2.4.

5.17 Prove Theorem 2.5. (Hint: use 1.6 and uniform continuity.)

5.18 Prove Theorem 2.6. (Hint: use 1.6. Given any partition $P = (x_0, \ldots, x_n)$, if f is non-decreasing, then let $f_1(x) = f(x_{i-1})$ for $x \in (x_{i-1}, x_i)$ and $f_2(x) = f(x_i)$ for $x \in (x_{i-1}, x_i)$.)

5.19 Consider the function

$$f(x) = \begin{cases} \sin(1/x) & \text{if } x \neq 0 \\ 0 & \text{if } x = 0. \end{cases}$$

Is f piecewise continuous on $[-1,1]$? Is f piecewise monotone on $[-1,1]$?

5.20 Invent a function which is monotone on $[0,1]$, but is not piecewise continuous.

5.21 (a) Show that for any sets A and B, $\chi_{A \cap B} = \chi_A \chi_B$.
     (b) Find similar expressions for $\chi_{A \cup B}$ and $\chi_{A \setminus B}$.
     (c) Show that $\chi_A + \chi_B = \chi_{A \cup B} + \chi_{A \cap B}$.

5.22 Show that χ_Q is discontinuous at every point.

5.23 Show that the function g in Example 3.3 is continuous at every irrational and discontinuous at every rational. (Hint: for x irrational, given $\epsilon > 0$, there are only finitely many rationals $p/q \in [x - 1, x + 1]$ with $1/q \geqslant \epsilon$. (Why?) Let $\delta > 0$ be so small that no such rational is in $[x - \delta, x + \delta]$.)

5.24 Prove that the function g of Example 3.3 is Riemann integrable on $[0,1]$. What is the value of $\int_0^1 g(x)dx$? (Hint: Given $\epsilon > 0$, let $f_1(x) = 0$ for all $x \in [0,1]$. If $1/n < \epsilon$, then $f_2(x) = 1/n$ except at those finitely many rationals $p/q \in [0,1]$ (written in "lowest terms") where $q < n$.)

5.25 Show that the class of Riemann integrable functions on $[a,b]$ is closed under uniform limits (i.e., if $\{f_n\}$ converges to f uniformly, and each f_n is Riemann integrable on $[a,b]$, then f is Riemann integrable on $[a,b]$.) Also show that in this case, $\int_a^b f(x)dx = \lim_{n \to \infty} \int_a^b f_n(x)dx$.

CHAPTER 2

Measurable Sets

6. Introduction

In his publications beginning in 1902, Henri Lebesgue presented one of the most exciting new ideas in the history of analysis. Some of his ideas had been anticipated by Borel and Cantor, but it was Lebesgue who fully developed the theory now known as Lebesgue measure and integration. His idea basically was to eliminate the deficiencies of the Riemann integral by starting with a partition of the range of f rather than a partition of the domain as in the Riemann integral.*

For example, if f is a bounded non-negative function defined on $[a,b]$, and if M is a strict upper bound of the values of $f(x)$, for $x \in [a,b]$, then the area under the graph of f could be approximated in the following

---

*In his own words (Henri Lebesgue, *Measure and the Integral,* Holden Day, San Francisco, 1966, p. 180): "It is clear . . . that we must break up not (a,b), but the interval $(f, \bar{f})$ bounded by the lower and upper bounds of $f(x)$ in (a,b). Let us do this with the aid of numbers y_i differing among themselves by less than ϵ. We are led to consider the values of $f(x)$ defined by $y_i \leqslant f(x) \leqslant y_{i-1}$. The corresponding values of x form a set E_i. . . . With some continuous functions it might consist of an infinity of intervals. For an arbitrary function it might be very complicated. But this matters little."

distinctly non-Riemann fashion. Let $(y_0, y_1, \ldots, y_n)$ be a partition of $[0, M]$ along the y-axis, and choose $y_i^* \in [y_{i-1}, y_i]$, for $i = 1, \ldots, n$.

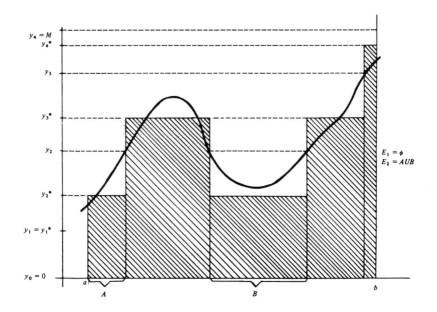

Define

$$E_i = \{x \in [a,b] \,|\, y_{i-1} \leqslant f(x) < y_i\}.$$

Then if $\ell(E_i)$ is the "length of E_i," an approximation to the area under the graph of f is given by the "Lebesgue sum"

$$y_1 * \ell(E_1) + y_2 * \ell(E_2) + \ldots + y_n * \ell(E_n).$$

In the end, of course, the Lebesgue measure will be the "limit" of such sums.

Lebesgue himself points out an analogy which contrasts his method with Riemann's in the simplest terms. Given a row of coins consisting, in order, of:

quarter, dime, penny, quarter, nickel, dime, dime,

Riemann would determine the total value of money by adding

$$25 + 10 + 1 + 25 + 5 + 10 + 10.$$

Lebesgue, on the other hand, would first count, for each denomination, the number of coins with that value. His sum would be

$$25 \cdot 2 + 10 \cdot 3 + 5 \cdot 1 + 1 \cdot 1.$$

Of course, the result is the same. For non-step functions, however, it is not so evident that Lebesgue's method gives the same result as Riemann's.

The difficulty with Lesbegue's procedure, as outlined above, is that it requires a systematic way of assigning "length" to subsets of the real line. In the diagram above, for example, one must be able to arrive at a "length" (or *measure*) for the set E_2. If E_2 turns out to be an interval or a union of intervals, there is no problem. But this need not be the case. Consider χ_Q, the characteristic function of the rationals, and any partition $(y_0, \ldots, y_n)$ of an interval containing its range $\{0,1\}$. At most two of the sets $E_i = \{x \in [0,1] \,|\, y_{i-1} \leqslant \chi_Q(x) < y_i\}$ will be non-empty (namely, those for which $y_{i-1} \leqslant 0 < y_i$ or $y_{i-1} \leqslant 1 < y_i$). If $y_{i-1} \leqslant 0 < y_i \leqslant 1$, then $E_i =$ the irrationals in $[0,1]$, and if $0 < y_{j-1} \leqslant 1 < y_j$, then $E_j =$ the rationals in $[0,1]$. (What happens if $y_{k-1} \leqslant 0 < 1 < y_k$?) What "length" can we reasonably assign to E_i and E_j? With a little imagination one can conceive of functions which yield even more complicated sets than $Q \cap [0,1]$ and its complement. We must have a method of assigning a "length" or measure to such sets, and any such assignment should "behave reasonably." Experience defines this reasonable behavior as follows.

6.1 Definition: Let S be a collection of subsets of $[a,b]$ which is closed under countable unions, that is, if $A_1, A_2, \ldots \in S$, then $\bigcup\limits_{n=1}^{\infty} A_n \in S$.

(i) A *set function* on S is a function which assigns to each set $S \in S$ a real number.

(ii) A set function μ on S is called a *measure* if
  (a) $0 \leqslant \mu(A) \leqslant b - a$ for every $A \in S$,
  (b) $\mu(\emptyset) = 0$,
  (c) whenever $A \subset B$ and $A, B \in S$, then $\mu(A) \leqslant \mu(B)$ (monotonicity),
  (d) whenever $A = \bigcup\limits_{n=1}^{\infty} A_n$, where $A_n \in S$ for $n = 1, 2, \ldots$ and $A_n \cap A_m = \emptyset$ for $n \neq m$, then

$$\mu(A) = \sum_{n=1}^{\infty} \mu(A_n)$$

(countable additivity).

We will also require that subintervals of $[a,b]$ be in S, and that the Lebesgue measure of an interval be equal to its ordinary length. Finally, we hope that every subset of $[a,b]$ will be in S; that is, we will be able to take the measure of any subset of $[a,b]$. However, we will see that this will be impossible if we wish to retain countable additivity (property (d) in the definition).

6.2 Example: Define $\mu(S)$ for any subset of $[0,1]$ by

$$\mu(S) = \begin{cases} 1 & \text{if } \frac{1}{2} \in S \\ \\ 0 & \text{if } \frac{1}{2} \notin S. \end{cases}$$

It is easy to verify that this defines a measure on $[0,1]$, but it is not consistent with our concept of the length of an interval since for example $\mu([\frac{1}{4},\frac{3}{4}]) = 1$ and $\mu([0,\frac{1}{3}]) = 0$.

6.3 Example: Define $\mu(S) = 0$ for every $S \subset [0,1]$. This is called the "trivial measure," and it is also inconsistent with lengths of intervals, although it obeys properties (a) – (d) of a measure.

6.4 Example: Anyone familiar with the Riemann integral might try to define a measure by

$$\mu(A) = \int_0^1 \chi_A(x)dx.$$

This is consistent with lengths of intervals, but as we have seen it fails to assign a measure to the set $Q \cap [0,1]$. In fact, if $S = \{A \subset [0,1] \,|\, \chi_A$ is Riemann integrable$\}$, then S is not closed under countable unions.

6.5 Example: The non-Riemann integrability of χ_Q results from the fact that for step functions $f_1 \leqslant \chi_Q \leqslant f_2$, we have

$$\int_0^1 f_1(x)dx \leqslant 0 < 1 \leqslant \int_0^1 f_2(x)dx.$$

(See Example 3.2.) We could instead consider only step functions f_2 lying above χ_Q. Or, in general if $A \subset [0,1]$, let

$$\mu(A) = \text{glb}\left\{\int_0^1 f_2(x)dx \,\Big|\, f_2 \text{ a step function and } f_2 \geqslant \chi_A \text{ on } [0,1]\right\}.$$

Clearly an equivalent expression is

$$\mu(A) = g\,\mathrm{lb}\{\sum_{i=1}^{n} (b_i - a_i) \,|\, A \subset \bigcup_{i=1}^{n} (a_i, b_i)\}.$$

This function is called "the Jordan content of A," and it is surely consistent with length for intervals. But it fails to be a measure; namely, $\mu(Q \cap [0,1])$ $= 1$, $\mu([0,1] \backslash Q) = 1$, but $\mu([0,1]) = 1$, contradicting property (d) of Definition 6.1. (See Exercise 9.8.)

Thus in this section we have seen the necessity of obtaining a concept of measure which obeys certain basic properties and agrees with our concept of the length of an interval. We have also seen that some obvious attempts to produce such a measure result in failure. In the next section we shall chart a course which results in a degree of success toward this end.

7. Outer Measure

For the remainder of this chapter, we will denote the unit interval $[0,1]$ by E. We will restrict our attention to subsets of E in order to avoid unnecessary complications. After we have developed the theory of measure on subsets of E, we will consider subsets of an arbitrary closed interval $[a,b]$, and even unbounded sets.

We begin with the following natural definition of measure for intervals, which will at the outset guarantee that our measure will be consistent with length for intervals.

7.1 Definition: The *outer measure* of any interval (open, closed, half-open) in E with endpoints $a < b$ will be the positive real number $b - a$.

We will use the symbol m^* to denote the outer measure function, so that $m^*([a,b]) = b - a$, for example. Note that it is possible to have $A \subsetneq B \subset E$, but $m^*(A) = m^*(B)$. This *outer* measure is our first attempt to define a measure. We will need to refine our ideas slightly to succeed in fulfilling this goal.

Now we extend our outer measure from intervals to arbitrary *open* subsets of E. This procedure depends on a general theorem about open sets of real numbers. Recall that any open set in $\mathfrak{R}$ can be written as a union of open intervals. The following theorem says even more.

7.2 Theorem: Every non-empty open set $G \subset \Re$ can be expressed uniquely as a finite or countably infinite union of pairwise disjoint open intervals.

Proof: Suppose first that G is bounded. Since G is open, for each $x \in G$ there is an open subinterval of G containing x.

Let

$$b_x = \mathrm{lub}\ \{y \,|\, (x,y) \subset G\}, \text{ and}$$

$$a_x = \mathrm{glb}\ \{z \,|\, (z,x) \subset G\}.$$

Let $I_x = (a_x, b_x)$, called the *component* of x in G. Clearly $x \in I_x$.

Now $I_x \subset G$, for if $w \in I_x$, say $x < w < b_x$, then by definition of b_x, there is a number y such that $w < y$ and $(x,y) \subset G$. Hence $w \in G$. The case where $a_x < w < x$ is handled similarly. (What about $w = x$?)

Also $a_x \notin G$ and $b_x \notin G$ (see Exercise 9.10).

Now we show that G can be expressed as the disjoint union of I_x's. Clearly $G = \underset{x \in G}{\cup} I_x$. Furthermore, for $x, y \in G$, either I_x and I_y are disjoint or they are identical. Indeed, suppose $I_x \cap I_y \neq \emptyset$. Then $a_y < b_x$ and $a_x < b_y$ (draw pictures to verify). But since $a_x \notin G$ and $I_y \subset G$, it follows that $a_x \notin I_y$; therefore $a_x \leqslant a_y$. Similarly $a_y \notin I_x$, so that $a_y \leqslant a_x$. Thus $a_x = a_y$. A similar argument shows that $b_x = b_y$, so that $I_x = I_y$. Hence any two intervals in the collection $\{I_x \,|\, x \in G\}$ are equal or disjoint, and G is the union of a disjoint collection of open intervals of the form I_x.

That this collection is countable follows from the fact that one can choose a rational number in each I_x (using the axiom of choice), and the disjointness of the intervals guarantees that no duplication will occur in the choice of rationals. The subset of Q thus chosen is countable, and is in one-to-one correspondence with the collection $\{I_x \,|\, x \in G\}$.

The proof that this representation of G is unique is left to the reader (Exercise 9.11).

The case where G is unbounded is handled in the same way, except that G may equal $(-\infty, \infty)$, which is trivial, or some components may be of the form $(-\infty, b)$ or (a, ∞). $\square$

Whenever we use open sets in the unit interval E, we shall use "open" to mean open in the relative topology of E as a subspace of $\Re$. That is, G is open in E if and only if $G = E$ intersected with some open subset of $\mathcal{R}$. Then it is clear that the theorem holds in the relative topology of E, in the sense that if G is open in E, there is a unique representation $G = \bigcup_i I_i$, where the I_i are disjoint intervals of the form $(a_i, b_i) \cap E$. For example, $(\frac{1}{2}, 1] = (\frac{1}{2}, 2) \cap E$. To avoid notational difficulties, when we write

$$G = \cup (a_i, b_i), \qquad G \subset E,$$

we allow the possibility that some (a_i, b_i) may be of the form $[0, b), (a, 1]$, or $[0, 1]$.

We now use the theorem to extend our definition of outer measure to open sets in E.

7.3 Definition: The outer measure $m^*(G)$ of an open set $G \subset E$ is defined as the real number $\sum_i (b_i - a_i)$, where $G = \bigcup_i (a_i, b_i)$ as in Theorem 7.2.

Since the theorem says that G is *uniquely* represented as a disjoint union of countably many intervals, this definition is unambiguous. Also, the number $\sum_{i=1}^{\infty} (b_i - a_i)$ exists since it is the sum of a series of positive terms with bounded partial sums. Finally, the order in which the terms of the series appear is unimportant since convergence is absolute. Note that $\sum_i (b_i - a_i)$ may be a finite sum if $\bigcup_i (a_i, b_i)$ is a finite union.

We next extend the definition of outer measure to *all* subsets of E, by approximating arbitrary sets with open sets.

7.4 Definition: The outer measure $m^*(A)$ of any set $A \subset E$ is defined to be the real number

$$\text{glb } \{m^*(G) | A \subset G \text{ and } G \text{ open in } E\}.$$

Clearly $m^*(A)$ exists for any $A \subset E$ since the set $\{m^*(G) | A \subset G$ and G open in $E\}$ is bounded below by 0 and thus has a finite greatest lower bound. It is also clear that if A is open in E, then the previous definition of $m^*(A)$ for open sets (Definition 7.3) agrees with this definition. In case A is a non-open interval, it is left as an exercise (Exercise 9.16) to show that this new definition agrees with the earlier definition (Definition 7.1) for intervals.

We will often use the following proposition, which is an immediate consequence of the definition of outer measure for arbitrary sets.

7.5 Proposition: Given any $\epsilon > 0$ and any set $A \subset E$, there exists an open set $G \subset E$ such that $A \subset G$ and $m^*(G) < m^*(A) + \epsilon$.

It would appear that we have arrived at a consistent manner of assigning a measure to any subset of E and that we need only verify that the desired properties for a measure hold. Unfortunately the outer measure we have defined fails to be countably additive on the class of all subsets of E. We will present an important example to show this, but first we need the following easy result.

7.6 Lemma: Let $A \subset E$. Then for any x, $m^*(A) = m^*(x + A)$, where $x + A = \{x + a \mid a \in A\}$ is called the translate of the set A by x. (Since we are restricted to subsets of E, we may need to translate modulo 1; for example, $[\frac{1}{2}, 1] + \frac{1}{4} = [\frac{3}{4}, 1] \cup [0, \frac{1}{4}]$.)

Proof: If A is an interval or a countable union of pairwise disjoint open intervals, the lemma is clearly true. Thus the lemma holds if A is any open set in E. For arbitrary $A \subset E$,

$$m^*(x + A) = \text{glb} \; \{m^*(G) \mid x + A \subset G \text{ and } G \text{ open in } E\}$$

$$= \text{glb} \; \{m^*(-x + G) \mid A \subset -x + G \text{ and } -x + G \text{ open in } E\}$$

$$\geqslant m^*(A).$$

The proof that $m^*(A) \geqslant m^*(x + A)$ is similar. $\square$

7.7 Example: Outer measure m^* is not countably additive on the class of all subsets of E. That is, there exist pairwise disjoint sets V_n that $E = \bigcup_{n=1}^{\infty} V_n$, but

$$1 = m^*(E) \neq \sum_{n=1}^{\infty} m^*(V_n).$$

To obtain V_n let $E_\alpha = \{x \in E \mid x - \alpha \text{ is rational}\}$ for each $\alpha \in E$. That is, E_α consists of all members of E which differ from α by a rational. Note that each E_α is countably infinite and that if $E_\alpha \cap E_\beta \neq \emptyset$, then $E_\alpha = E_\beta$ (the reader should verify this; see Exercise 9.21).

Using the axiom of choice, let V be a set consisting of exactly one x_α from each distinct set E_α. Thus, no two distinct members of V can differ by a rational. Now let $q_1, q_2, \ldots$ be an enumeration of the rational numbers in E, and for each n, let $V_n = q_n + V$.

Claim 1: $V_n \cap V_m = \emptyset$ for $n \neq m$.

Proof: Let $y \in V_n \cap V_m$. Then there exist $x_\alpha, x_\beta \in V$ such that $x_\alpha + q_n = y = x_\beta + q_m$. Then $x_\alpha - x_\beta$ must be rational, which means that $x_\alpha = x_\beta$. Hence $q_n = q_m$ and $V_n = V_m$.

Claim 2: $E = \bigcup\limits_{n=1}^{\infty} V_n$.

Proof: If $x \in E$, then $x \in E_\alpha$ for some α. But there is one representative of E_α in V, say x_α, so that $x = x_\alpha + q_n$ for some rational q_n. That is, $x \in V_n$.

Claim 3: m^* is not countably additive on E.

Proof: Clearly $1 = m^*(E) = m^*(\bigcup\limits_{n=1}^{\infty} V_n)$. By the lemma (7.6), $m^*(V) = m^*(V_n)$ for all $n = 1, 2, \ldots$, since the V_n's are translates of V. Now by Claim 1, the V_n's are disjoint, so if countable additivity were to hold, we would have $1 = \sum\limits_{n=1}^{\infty} m^*(V_n) = \sum\limits_{n=1}^{\infty} m^*(V)$. That is, we would have infinitely many equal numbers $m^*(V)$ adding up to 1. This is clearly impossible. □

This example would suggest that we have failed in our attempt to define a measure. However, the amount of work necessary to produce the example, and the use of the axiom of choice, suggest that perhaps all is not lost if we are willing to settle for slightly less than our original goal. To ensure that the resulting Lebesgue Integral satisfies desired properties, we insist on retaining countable additivity as a key property for our measure. However, we are forced to relax our hope that *every* subset of E have a measure assigned to it. We therefore set out to find a large class of subsets of E on which m^* will be countably additive. There are many ways to do this and not all methods result in the same class of sets. In section 8 we will present one method of obtaining such a class of sets (called Lebesgue measurable sets).

8. Measurable Sets

In order to produce a large class of sets on which m^* will be countably additive, we need the following definition.

8.1 Definition: The inner measure $m_*(A)$ of any subset $A \subset E$ is defined to be the number $1 - m^*(E \backslash A)$.

It is clear that $m_*(A)$ exists and is non-negative for each $A \subset E$ since $m^*(E \backslash A)$ exists and is between 0 and 1.

We are now in a position to give the definition of measurable set. We will eventually show that the class M of all measurable sets is a large enough collection of subsets of E for our purposes, and that m^* is a measure (in particular, m^* is countably additive) on M.

8.2 Definition: A set $A \subset E$ is said to be (Lebesgue) *measurable* if $m_*(A) = m^*(A)$. In this case, the *measure of A, denoted $m(A)$*, is the number $m_*(A) = m^*(A)$.

8.3 Proposition: A set $A \subset E$ is measurable if and only if

$$m^*(A) + m^*(E \backslash A) = 1.$$

Proof: This follows immediately from the two preceding definitions.
□

8.4 Corollary: A set $A \subset E$ is measurable if and only if $E \backslash A$, the complement of A in E, is measurable.

It is easy to show that $\emptyset$, E, and finite sets of points in E are measurable. Also, every interval in E is measurable (Exercise 9.22), and measure is consistent with length for intervals. It is also true that open sets are measurable, but we postpone proof of this until we have shown countable additivity of m^* on the class of measurable sets in Chapter Three.

We now present further examples of measurable sets, preceded by several lemmas which we will find very useful in proving countable additivity.

8.5 Lemma: (Monotonicity of m^* and m_*). Given $A \subset B \subset E$, we have

(1) $m^*(A) \leqslant m^*(B)$, and

(2) $m_*(A) \leqslant m_*(B)$.

Proof: (1) follows easily from the definition of outer measure (see Definition 7.4 and Exercise 9.17).

For (2), note that $E \backslash B \subset E \backslash A$, so that

$$m^*(E \backslash B) \leqslant m^*(E \backslash B)$$

by part (1). Thus

$$m_*(A) = 1 - m^*(E \backslash A) \leqslant 1 - m^*(E \backslash B) = m_*(B).$$

$\square$

8.6 Lemma: (Subadditivity of m^*). Suppose $\{A_n \mid n = 1,2, \ldots \}$ is a collection of subsets of E. Then

$$m^*(\bigcup_{n=1}^{\infty} A_n) \leqslant \sum_{n=1}^{\infty} m^*(A_n).$$

Proof: We present the proof in several parts.

(1) *Finite collections of open intervals.*

It is obvious that if I_1 and I_2 are open intervals in E, then

$$m^*(I_1 \cup I_2) \leqslant m^*(I_1) + m^*(I_2).$$

(In fact, equality holds if and only if $I_1 \cap I_2 = \emptyset$.) It follows that

$$m^*(\bigcup_{k=1}^{n} I_k) \leqslant \sum_{k=1}^{n} m^*(I_k)$$

for any finite collection of open intervals. (See Exercise 9.24.)

(2) *Countable collections of open intervals.*

Given $\bigcup_{n=1}^{\infty} I_n$, write this open set as $\bigcup_{n=1}^{\infty} J_n$, where the J_n's are disjoint open intervals (Theorem 7.2). Now given $\epsilon > 0$, there is an N such that

$$\sum_{n=N+1}^{\infty} m^*(J_n) < \epsilon. \quad \text{(Why?)}$$

Thus,

$$m^*(\bigcup_{n=1}^{\infty} I_n) = \sum_{n=1}^{\infty} m^*(J_n) < \sum_{n=1}^{N} m^*(J_n) + \epsilon.$$

For each $n = 1,2, \ldots, N$, there is an open interval $K_n = (a_n, b_n)$ such that $\overline{K}_n = [a_n, b_n] \subset J_n$ and such that

$$m^*(J_n) < m^*(K_n) + \epsilon/N.$$

Thus,

$$m^*(\bigcup_{n=1}^{\infty} I_n) < \sum_{n=1}^{N} m^*(K_n) + 2\epsilon.$$

Now $\bigcup_{n=1}^{N} \overline{K}_n$ is a compact (closed, bounded) set contained in the open open cover $\bigcup_{n=1}^{\infty} I_n$. By the definition of compactness, $\bigcup_{n=1}^{N} \overline{K}_n$ is contained in the union of some finite subcollection $\bigcup_{j=1}^{r} I_{n_j}$, so that $\bigcup_{n=1}^{N} \overline{K}_n \subset \bigcup_{j=1}^{r} I_{n_j}$. Therefore, $\sum_{n=1}^{N} m^*(K_n) \leqslant m^*(\bigcup_{j=1}^{r} I_{n_j}) \leqslant \sum_{j=1}^{r} m^*(I_{n_j})$ by part (1). Therefore we have

$$m^*(\bigcup_{n=1}^{\infty} I_n) \leqslant \sum_{j=1}^{r} m^*(I_{n_j}) + 2\epsilon \leqslant \sum_{n=1}^{\infty} m^*(I_n) + 2\epsilon.$$

Since ϵ was arbitrary, the lemma holds for countable collections of open intervals.

(3) *Countable collections of arbitrary sets.*

Let $\{A_n \mid n = 1,2, \dots \}$ be a collection of arbitrary subsets of E. Given $\epsilon > 0$, for each n, there exists an open set G_n such that $A_n \subset G_n$ and

$$m^*(G_n) < m^*(A_n) + \epsilon/2^n \qquad \text{(Proposition 7.5)}.$$

But G_n has a unique representation $\bigcup_k I_{n,k}$ as a union of pairwise disjoint open intervals, and by definition,

$$m^*(G_n) = \sum_k m^*(I_{n,k}) < m^*(A_n) + \epsilon/2^n.$$

Therefore, $\bigcup_{n=1}^{\infty} A_n \subset \bigcup_{n=1}^{\infty} \bigcup_k I_{n,k}$ and

$$m^*(\bigcup_{n=1}^{\infty} A_n) \leqslant m^*(\bigcup_{n=1}^{\infty} \bigcup_k I_{n,k}) \leqslant \sum_{n=1}^{\infty} \sum_k m^*(I_{n,k}) \leqslant \sum_{n=1}^{\infty} m^*(A_n) + \epsilon.$$

Since ϵ was arbitrary, the lemma follows.

(Note that $\sum_{n=1}^{\infty} m^*(A_n)$ may be infinite but $m^*(\bigcup_{n=1}^{\infty} A_n)$ must be finite since $\bigcup_{n=1}^{\infty} A_n \subset E$.) $\qquad\qquad\square$

8.7 Lemma: For any set $A \subset E$, $m_*(A) \leqslant m^*(A)$.

    Proof: Clearly $A \cup (E \backslash A) = E$, so that $m^*(A \cup (E \backslash A)) = 1$. But by Lemma 8.6,

$$m^*(A \cup (E \backslash A)) \leqslant m^*(A) + m^*(E \backslash A).$$

Therefore,

$$m^*(A) \geqslant 1 - m^*(E \backslash A) = m_*(A).$$

    □

8.8 Example: Any set A with outer measure $m^*(A) = 0$ is measurable and has measure 0 since

$$0 = m^*(A) \geqslant m_*(A) \geqslant 0 \text{ implies } m^*(A) = m_*(A) = 0.$$

    (1) The empty set $\emptyset$ and any finite set of points $\{p_1, \ldots, p_n\} \subset E$ are measurable sets with measure 0. We need only show that $m^*(\{p_1, \ldots, p_n\}) = 0$, but this is true since for any $\epsilon > 0$, the open set

$$G = (p_1 - \epsilon/n, \ p_1 + \epsilon/n) \cup (p_2 - \epsilon/n, \ p_2 + \epsilon/n) \cup \ldots$$

$$\cup (p_n - \epsilon/n, \ p_n + \epsilon/n)$$

has outer measure $\leqslant 2\epsilon$. By definition,

$$m^*(\{p_1, \ldots p_n\}) \leqslant m^*(G) \leqslant 2\epsilon, \quad \text{so that}$$

$$m^*(\{p_1, \ldots, p_2\}) = 0.$$

    (2) Every countably infinite set $\{p_1, p_2, \ldots\} \subset E$ is measurable and has measure 0. (In particular, $m(Q \cap E) = 0$.) Again we show that the outer measure is 0. Indeed, by subadditivity (Lemma 8.6), and by part (1),

$$m^*(\{p_1, \ldots\}) \leqslant \sum_{i=1}^{\infty} m^*(\{p_i\}) = 0.$$

This example raises the question: are there any sets of measure 0 which are not countable? We will see that the answer is yes in the next chapter.

    (3) If $B \subset A \subset E$ and $m^*(A) = 0$, then by monotonicity of outer measure (Lemma 8.5), $m^*(B) = 0$, so that B is measurable, and has measure 0.

8.9 Example: Recall from Corollary 8.4 that the complement $E\backslash A$ of any measurable set A is measurable. It follows in this case that the measure of $E\backslash A$ is $1 - m^*(A)$.

(1) The complement of any countable subset of E is measurable and has measure 1. For example, the set of irrational numbers in E is measurable, with measure 1.

(2) Complements of intervals are measurable, since intervals are measurable.

In subsequent sections we will show that all open (and hence all closed) subsets of E are measurable. We will also prove that outer measure m^* is a measure, satisfying countable additivity, on the class of all measurable sets, and we will produce an example of a set of measure 0 which has uncountably many points.

9. Exercises

9.1 Let $f(x) = x^2$ on $[0,1]$, $y_0 = 0$, $y_1 = \frac{1}{4}$, $y_2 = \frac{1}{2}$, $y_3 = 1$, $y_4 = 3$. Find E_1, E_2, E_3, E_4.

9.2 Prove that if μ is a measure on S and A, $B \in S$, and $A \cap B = \emptyset$, then $\mu(A \cup B) = \mu(A) + \mu(B)$.

9.3 Prove that if μ is a measure on S and $\{x\} \in S$ for every $x \in [a,b]$, and $\mu(\{x\}) = \mu(\{y\})$ for all $x, y \in [a,b]$, then $\mu(Q) = 0$.

9.4 Let S be a collection of subsets of $[a,b]$ which is closed under countable unions and under complements in $[a,b]$ (that is, if $A \in S$, then $[a,b]\backslash A \in S$).
(a)　Prove that S is closed under countable intersections.
(b)　Prove that if A, $B \in S$, then $A\backslash B \in S$.
(c)　Show that if μ is a measure on S, then

$$\mu(A \cup B) = \mu(A) + \mu(B) - \mu(A \cap B).$$

(Hint for (c): $A \cup B = (A\backslash B) \cup (B\backslash A) \cup (A \cap B)$ disjointly.)

9.5 Prove that the set function μ of Example 6.2 is a measure.

9.6 Prove that the "trivial measure" of Example 6.3 is indeed a measure,

9.7 Show that the collection S of Example 6.4 is not closed under countable unions. Show however that μ satisfies (a), (b), (c) of Definition 6.1, and that μ is consistent with lengths of intervals.

9.8 (a) Show that the two expressions for μ in Example 6.5 are equivalent.

(b) Show that "Jordan content" is consistent with lengths of intervals.

(c) Show that $\mu(Q \cap [0,1]) = 1$ and $\mu([0,1] \setminus Q) = 1$.

(d) Prove or disprove each of the defining properties of measure for Jordan content.

9.9 Find the component of π in the open set $\bigcup_{n=1}^{\infty} (n, n + 1/n)$.

9.10 Prove that $a_x \notin G$, in the proof of Theorem 7.2.

9.11 Prove uniqueness in Theorem 7.2; if $G = \bigcup_n I_n = \bigcup_k J_k$, where I_n, J_k are open intervals and $I_n \cap I_m = \emptyset$, $J_k \cap J_\ell = \emptyset$ for all n, m, k, ℓ, then show that for every n there is a k such that $I_n = J_k$, and for every k there is an n such that $J_k = I_n$. (Hint: if $x \in I_n$, show that $I_n = I_x$.)

9.12 Find $m^*(G)$, where $G = (0,1] \setminus \{1/n \mid n = 1, 2, \ldots\}$.

9.13 Do there exist open subsets G_1, G_2 of E such that $G_1 \neq G_2$ but $m^*(G_1) = m^*(G_2)$?

9.14 Prove or disprove that if $G_1 \subsetneq G_2 \subset E$ are open, then $m^*(G_1) < m^*(G_2)$.

9.15 Prove that m^* is countably additive on the class of open subsets of E.

9.16 (a) Show that m^* is consistently defined for closed intervals by Definition 7.4; that is, if $[a,b] \subset E$, show that

$$b - a = \mathrm{glb}\,\{m^*(G) \mid [a,b] \subset G \text{ and } G \text{ open in } E\}.$$

(b) Is there an open set $G \subset E$ with $[1/2, 3/4] \subset G$ and $m^*(G) = \frac{1}{4}$?

9.17 Show that if $A \subset B \subset E$, then $m^*(A) \leqslant m^*(B)$. (Definition 7.4)

9.18 If $A \subset [0, \frac{1}{2}]$ and $B \subset (\frac{1}{2}, 1]$, show that $m^*(A \cup B) = m^*(A) + m^*(B)$.

9.19 Given $A \subset E$, show that $m^*(\{1 - x \mid x \in A\}) = m^*(A)$. (Hint: see proof of Lemma 7.6. Give details.)

9.20 Describe $E_{1/2}$ and $E_{\sqrt{2}-1}$ in Example 7.7.

9.21 In Example 7.7, prove that if $E_\alpha \cap E_\beta \neq \emptyset$, then $E_\alpha = E_\beta$.

9.22 Prove from Definition 8.2 that all subintervals of E are measurable.

9.23 (a) If $A \subset E$ is measurable, show that $x + A$ is measurable.

(b) If $A \subset E$ is measurable, show that $\{1 - x \mid x \in A\}$ is measurable.

9.24 (a) Show that $m^*(I_1 \cup I_2) \leqslant m^*(I_1) + m^*(I_2)$ for I_1, I_2 open intervals in E.

(b) Show that $m^*(\bigcup_{k=1}^{n} I_k) \leqslant \sum_{k=1}^{n} m^*(I_k)$ for open intervals I_k in E.

9.25 Referring to Lemma 8.6, give an example of a collection $\{A_n\}_{n=1}^{\infty}$ of subsets of E such that

$$m^*(\bigcup_{n=1}^{\infty} A_n) < \sum_{n=1}^{\infty} m^*(A_n).$$

9.26 Prove that $m^*(A) = \text{glb} \{\sum_{n=1}^{\infty} m^*(J_n)| J_n \text{ open intervals and } A \subset \bigcup_{n=1}^{\infty} J_n\}$.
(Note that the J_n need not be disjoint. Use Lemma 8.6.)

9.27 If $m^*(A) + m^*(B) = m_*(A) + m_*(B)$, prove that A is measurable and B is measurable. (Hint: Lemma 8.7.)

9.28 Example 8.8(2) implies the surprising result that for any $\epsilon > 0$, the dense set $Q \cap E$ is contained in an open set with outer measure less than ϵ. Write down an expression for such an open set G. (Hint: let $Q \cap E = \{q_1, q_2, \ldots\}$. Let $G = \bigcup_{n=1}^{\infty} I_n$, where $q_n \in I_n$ for each n, and I_n is a specified open interval. See the proof of Lemma 8.6(3) and Example 8.8(1) for an idea of what I_n should be.)

9.29 Show that m^* is not (finitely) additive on the collection of all subsets of E. (See Lemma 8.6 and Example 7.7.)

CHAPTER 3

Properties of Measurable Sets

10. Countable Additivity

In this section we will show that outer measure m^* is countably additive when restricted to the class of measurable subsets of E. We first need two lemmas.

10.1 Lemma: If G_1 and G_2 are open subsets of E, then

$$m^*(G_1) + m^*(G_2) \geqslant m^*(G_1 \cup G_2) + m^*(G_1 \cap G_2).$$

Proof: If G_1 and G_2 are intervals, the Lemma is obvious. If G_1 and G_2 are both finite unions of open intervals, the proof of the Lemma is contained in Exercise 16.1 at the end of the Chapter.

For G_1 and G_2 arbitrary open subsets of E, we know that $G_1 = \overset{\infty}{\underset{i=1}{\cup}} I_i$ and $G_2 = \overset{\infty}{\underset{j=1}{\cup}} J_j$, where the I_i are disjoint open intervals and the J_j are disjoint open intervals. Given $\epsilon > 0$, there is an integer N such that $\overset{\infty}{\underset{i=N+1}{\Sigma}} m^*(I_i) < \epsilon$ and $\overset{\infty}{\underset{j=N+1}{\Sigma}} m^*(J_j) < \epsilon$ (Why?).

Define $G_1' = \overset{N}{\underset{i=1}{\cup}} I_i$ and $G_2' = \overset{n}{\underset{j=1}{\cup}} J_j$. Also define $G_1'' = \overset{\infty}{\underset{i=N+1}{\cup}} I_i$ and $G_2'' = \overset{\infty}{\underset{j=N+1}{\cup}} J_j$. Then $G_1 \cup G_2 = G_1' \cup G_2' \cup G_1'' \cup G_2''$ so that $m^*(G_1 \cup G_2) \leqslant m^*(G_1' \cup G_2') + 2\epsilon$ by subadditivity (Lemma 8.6).

Similarly $G_1 \cap G_2 = (G_1' \cup G_1'') \cap (G_2' \cup G_2'') \subset (G_1' \cap G_2') \cup G_1'' \cup G_2''$. Thus $m^*(G_1 \cap G_2) \leqslant m^*(G_1' \cap G_2') + 2\epsilon$.

Now $m^*(G_1) + m^*(G_2) \geqslant m^*(G_1') + m^*(G_2') = m^*(G_1' \cup G_2')$ $+ m^*(G_1' \cap G_2')$ since G_1' and G_2' are finite unions of open intervals. Thus $m^*(G_1) + m^*(G_2) \geqslant m^*(G_1 \cup G_2) - 2\epsilon + m^*(G_1 \cap G_2) - 2\epsilon$ from above. Since ϵ was arbitrary, the Lemma follows. □

10.2 Lemma: If A_1 and A_2 are subsets of E, then

(1) $m^*(A_1) + m^*(A_2) \geqslant m^*(A_1 \cup A_2) + m^*(A_1 \cap A_2)$ and
(2) $m_*(A_1) + m_*(A_2) \leqslant m_*(A_1 \cup A_2) + m_*(A_1 \cap A_2)$.

Proof: Given $\epsilon > 0$, there exist open sets $G_1, G_2 \subset E$ such that $G_1 \supset A_1$, $G_2 \supset A_2$, and $m^*(G_i) < m^*(A_i) + \epsilon$ for $i = 1,2$. Thus $m^*(A_1 \cup A_2) \underset{8.5}{\leqslant} m^*(G_1 \cup G_2) \underset{10.1}{\leqslant} m^*(G_1) + m^*(G_2) - m^*(G_1 \cap G_2)$ $\leqslant m^*(A_1) + \epsilon + m^*(A_2) + \epsilon - m^*(A_1 \cap A_2)$. Since this is true for all $\epsilon > 0$, the result in (1) follows.

To prove part (2), use $E \backslash A_1$ and $E \backslash A_2$ in place of A_1 and A_2 above (Exercise 16.2). □

10.3 Corollary: If A_1 and A_2 are measurable subsets of E, then $A_1 \cup A_2$ and $A_1 \cap A_2$ are measurable.

Proof: By the Lemma, $m^*(A_1) + m^*(A_2) \geqslant m^*(A_1 \cup A_2)$ $+ m^*(A_1 \cap A_2) \underset{8.7}{\geqslant} m_*(A_1 \cup A_2) + m_*(A_1 \cap A_2) \geqslant m_*(A_1) + m_*(A_2)$. But $m^*(A_i) = m_*(A_i)$ by hypothesis, so the inequalities become equal signs. Now $m^*(A_1 \cup A_2) + m^*(A_1 \cap A_2) = m_*(A_1 \cup A_2) + m_*(A_1 \cap A_2)$ together with Lemma 8.7 imply that $m^*(A_1 \cup A_2) = m_*(A_1 \cup A_2)$ and $m^*(A_1 \cap A_2) = m_*(A_1 \cap A_2)$. (Exercise 9.27.) □

10.4 Corollary: If $A_1 \cap A_2 = \phi$ for sets A_1, $A_2 \subset E$, then

$$m_*(A_1 \cup A_2) \geqslant m_*(A_1) + m_*(A_2).$$

Proof: Since $m^*(\phi) = 0$, this is obvious from the Lemma. □

10.5 Corollary: If $\{A_i\}_{i=1}^{\infty}$ are pairwise disjoint measurable subsets of E, then

$$m_*(\bigcup_{i=1}^{\infty} A_i) \geqslant \sum_{i=1}^{\infty} m_*(A_i).$$

Proof: For any integer N, $m_*(\overset{\infty}{\underset{i=1}{\cup}} A_i) \underset{8.5}{\geqslant} m_*(\overset{N}{\underset{i=1}{\cup}} A_i) \geqslant \overset{N}{\underset{i=1}{\Sigma}} m_*(A_i)$ by the previous Corollary extended to finite unions by induction. Letting $N \to \infty$, we obtain $m_*(\overset{\infty}{\underset{i=1}{\cup}} A_i) \geqslant \overset{\infty}{\underset{i=1}{\Sigma}} m_*(A_i)$. $\square$

We are now in a position to prove the major theorem of this section. Once we know that m^* is countably additive on the measurable subsets of E, then we will know that m^* is a measure according to our earlier definition (Definition 6.1).

10.6 Theorem: (Countable Additivity). Let $\{A_i\}_{i=1}^{\infty}$ be a pairwise disjoint collection of measurable sets in E. Then $\overset{\infty}{\underset{i=1}{\cup}} A_i$ is measurable and $m^*(\overset{\infty}{\underset{i=1}{\cup}} A_i) = \overset{\infty}{\underset{i=1}{\Sigma}} m^*(A_i)$.

Proof: Since each A_i is measurable, $m^*(A_i) = m_*(A_i)$ for each i. Thus $m^*(\overset{\infty}{\underset{i=1}{\cup}} A_i) \underset{8.6}{\leqslant} \overset{\infty}{\underset{i=1}{\Sigma}} m^*(A_i) = \overset{\infty}{\underset{i=1}{\Sigma}} m_*(A_i) \underset{10.5}{\leqslant} m_*(\overset{\infty}{\underset{i=1}{\cup}} A_i)$. By Lemma 8.7, $m_*(\overset{\infty}{\underset{i=1}{\cup}} A_i) \leqslant m^*(\overset{\infty}{\underset{i=1}{\cup}} A_i)$, so $\overset{\infty}{\underset{i=1}{\cup}} A_i$ is measurable and $m^*(\overset{\infty}{\underset{i=1}{\cup}} A_i) = \overset{\infty}{\underset{i=1}{\Sigma}} m^*(A_i)$. $\square$

There are many useful results which follow easily from countable additivity. We list several as corollaries.

10.7 Corollary: All open and closed subsets of E are measurable.

Proof: Open sets are countable pairwise disjoint unions of open intervals, which are measurable. Closed sets are complements of open sets. $\square$

10.8 Corollary: There exist non-measurable sets.

Proof: In Example 7.7 we produced pairwise disjoint sets V_n all of equal outer measure. Since $\overset{\infty}{\underset{m=1}{\cup}} V_n = E$, if the V_n were measurable sets, $\overset{\infty}{\underset{n=1}{\Sigma}} m^*(V_n)$ would be 1. This is clearly impossible, so the V_n's are not measurable. That is, $m_*(V_n) < m^*(V_n)$ for each of these sets. $\square$

10.9 Corollary: If $A_1 \subset A_2 \subset A_3 \subset \cdots$ are measurable subsets of E, then $\overset{\infty}{\underset{i=1}{\cup}} A_i$ is measurable and $m^*(\overset{\infty}{\underset{i=1}{\cup}} A_i) = \underset{i \to \infty}{\lim} m^*(A_i)$.

Proof: Consider the sequence $A_1, A_2 \backslash A_1, A_3 \backslash A_2, \cdots$. Each set is measurable by Corollary 10.3, since $A_n \backslash A_{n-1} = A_n \cap (E \backslash A_{n-1})$. Furthermore, the sets are pairwise disjoint, so $\overset{\infty}{\underset{i=1}{\cup}} A_i = A_1 \cup (\overset{\infty}{\underset{i=2}{\cup}} [A_i \backslash A_{i-1}])$ is measurable by countable additivity. Also,

$$m^*(\overset{\infty}{\underset{i=1}{\cup}} A_i) = m^*(A_1) + \overset{\infty}{\underset{i=2}{\Sigma}} m^*(A_i \backslash A_{i-1}) = m^*(A_1) + \lim_{N \to \infty} \overset{N}{\underset{i=2}{\Sigma}} m^*(A_i \backslash A_{i-1})$$

$$= \lim_{N \to \infty} (m^*(A_1 \cup \overset{n}{\underset{i=2}{\cup}} [A_i \backslash A_{i-1}])) = \lim_{N \to \infty} (m^*(A_N)). \qquad \square$$

10.10 Corollary: If $A_1 \supset A_2 \supset A_3 \supset \cdots$ are measurable subsets of E, then $\overset{\infty}{\underset{i=1}{\cap}} A_i$ is measurable and $m^*(\overset{\infty}{\underset{i=1}{\cap}} A_i) = \lim_{i \to \infty} m^*(A_i)$.

Proof: Exercise 16.7. $\qquad \square$

10.11 Corollary: Let $\{A_i\}_{i=1}^{\infty}$ be any (not necessarily disjoint) countable collection of measurable sets in E. Then

(1) $\overset{\infty}{\underset{i=1}{\cup}} A_i$ is measurable and

(2) $\overset{\infty}{\underset{i=1}{\cap}} A_i$ is measurable.

Proof: (1) Write $\overset{\infty}{\underset{i=1}{\cup}} A_i$ as the union of pairwise disjoint sets $A_1 \cup (A_2 \backslash A_1) \cup (A_3 \backslash [A_1 \cup A_2]) \cup \cdots \cup (A_n \backslash [\overset{n-1}{\underset{i=1}{\cup}} A_i]) \cup \cdots$. Each set $A_n \backslash [\overset{n-1}{\underset{i=1}{\cup}} A_i]$ is measurable by Corollary 10.3 and induction (Exercise 16.5). The result follows from countable additivity (Theorem 10.6).

(2) Write $\overset{\infty}{\underset{i=1}{\cap}} A_i = E \backslash (\overset{\infty}{\underset{i=1}{\cup}} [E \backslash A_i])$ and use part (1). $\qquad \square$

11. Summary

Let us pause to take account of our accomplishments. Our original (ambitious) goal was to construct a set function m defined for all subsets of $E = [0,1]$ and satisfying

(1) $0 \leqslant m(A) \leqslant 1$ for any $A \subset E$.
(2) If $A \subset B$, then $m(A) \leqslant m(B)$ for $A, B \subset E$.
(3) $m(\phi) = 0$ and $m(E) = 1$.
(4) $m(\overset{\infty}{\underset{i=1}{\cup}} A_i) = \overset{\infty}{\underset{i=1}{\Sigma}} m(A_i)$ for pairwise disjoint subsets A_i of E.

In addition we expected m to agree with our concept of length of an interval.

In Chapter 2, we built up a definition of the outer measure m^* based on certain of these properties. However, we learned that m^* was not countably additive (property 4 above) on the class of all subsets of E. Our solution was to retain m^* but to apply it only to certain (measurable) subsets of E. In section 10 we saw that this procedure resulted in a class of sets on which m^* obeys properties (1) - (4) of a measure. Let us state this fact formally.

11.1 Theorem: Let M be the class of measurable subsets of E. Then M contains all intervals in E and is closed under countable unions, countable intersections, and relative complements. Also, m^* is a measure on M. That is, m^* obeys properties (1) - (4) above when restricted to measurable subsets of E. Also m^* applied to any interval gives the length of the interval.

NOTATION: For the remainder of the text we will use $m(A)$ rather than $m^*(A)$ whenever A is a measurable set.

12.** Borel Sets and the Cantor Set

The most difficult part of our development has been proving countable additivity. A great deal of this difficulty stemmed from having to verify that various sets were measurable according to our definition of measurable set. One might reasonably ask whether M or some other large class of sets on which m^* is a measure could be more easily defined. We will present one such large class (the Borel Sets) and then show that the class M of measurable sets has many more sets than the class of Borel Sets. First we need the following concept.

12.1 Definition: Let S be a collection of subsets of E. Then S is called a *countably-additive class of sets* (or a *σ-algebra*) if

(1) $E \in S$.
(2) $A \in S \Rightarrow E \backslash A \in S$.
(3) $\bigcup_{i=1}^{\infty} A_i \in S$ whenever $\{A_i\}_{i=1}^{\infty} \subset S$.

The following result may easily be proved.

12.2 Proposition: If S is a countably-additive class of sets in E, then

(1) $\phi \in S$.

(2) $\bigcap_{i=1}^{\infty} A_i \in S$ whenever $\{A_i\}_{i=1}^{\infty} \subset S$.

(3) Finite unions and intersections of members of S are in S.

Proof: Exercise 16.10. $\square$

Two examples of countably-additive classes are the class of all sub-sets of E and the class of all measurable subsets of E (see Exercise 16.11).

12.3 Theorem: Let A be any class of subsets of E. Then there exists a minimal countably-additive class $S \supset A$ (Here "minimal" means that if W is any countably additive class containing A, then $W \supset S$).

Proof: Let S be the intersection of all countably-additive classes containing A. Since the class of all subsets of E contains A and is a countably-additive class, this intersection exists.

It is easily verified that S is a countably-additive class of sets containing A (Exercise 16.13). Since S is the intersection of all such classes it is clearly minimal. $\square$

12.4 Definition: Consider the minimal countably-additive class B which contains all open intervals in E. This class B is called the class of *Borel Sets* in E.

The previous theorem assures us that B exists. It is clear that B contains a vast number of sets (all open sets, closed sets, intervals, count-able unions and intersections of these, and many more). Since M is a countably-additive class containing the open intervals, we know that $B \subset M$. Could it be that $B = M$ and that we could have obtained M in this easier way? Or, even if $M \neq B$, are there enough additional sets in M to justify our previous labors? To answer these questions and provide one of the most famous examples in analysis, we present the Cantor Set.

12.5 Example: (The Cantor Set)

Let

$$C_1 = E\backslash \left(\frac{1}{3}, \frac{2}{3} \right)$$

$$C_2 = C_1 \backslash \left[\left(\frac{1}{9}, \frac{2}{9} \right) \cup \left(\frac{7}{9}, \frac{8}{9} \right) \right]$$

$$\vdots$$

$$C_n = C_{n-1} \setminus \left[\left(\frac{1}{3}, \frac{2}{3^n} \right) \cup \cdots \cup \left(\frac{3^n - 2}{3^n}, \frac{3^n - 1}{3^n} \right) \right] .$$

That is, C_n results from removing the open middle thirds of each interval in C_{n-1}.

12.6 Definition: The Cantor Set $C = \bigcap\limits_{n=1}^{\infty} C_n$.

12.7 Theorem: The Cantor Set is an uncountable measurable set with measure zero.

Proof: Clearly C is closed since its complement is open. Thus C is measurable. Now $m(C_n) = \frac{2}{3} m(C_{n-1}) = \frac{2}{3}(\frac{2}{3} m(C_{n-2})) = \cdots = (\frac{2}{3})^n m(E)$, and C is the intersection of a nested sequence of measurable sets C_n, so by Corollary 10.10 $m(C) = \lim\limits_{n \to \infty} C_n = \lim\limits_{n \to \infty} (\frac{2}{3})^n = 0$.

To show that C is uncountable, expand the numbers in E in ternary form (i.e. base three). (See Exercise 16.16.) Now C consists of the set of all points in E which have no 1 in their ternary expansion (e.g. $C_1 = $ all numbers which can be written in ternary form with a 0 or a 2 in the first place after the decimal point—note that $\frac{1}{3} = .02222 \cdots$ and $\frac{2}{3} = .2000 \cdots$ here). But the set of all sequences of 0's and 2's is uncountable by a familiar diagonal argument (Exercise 16.20). □

12.8 Corollary: Every subset of the Cantor set is measurable with measure zero.

Proof: All subsets of sets of measure zero are measurable with measure zero by 8.5 and 8.8. □

For those with a knowledge of cardinal arithmetic, the above Corollary shows that there are at least $2^c = 2^{2^{\aleph_0}}$ measurable sets in E. (The

Cantor Set has $2^{\aleph_0}$ points since it is in a one-to-one correspondence with
E.) Now the collection of Borel Sets is generated by operations (unions,
intersections, and complements) on countable collections of open intervals
and on countable collections of the resulting sets etc. It follows from a
rather involved argument using ordinal numbers that there are at most
$2^{\aleph_0} = c$ Borel Sets. Thus the Cantor Set alone contains $2^{2^{\aleph_0}}$ measurable
sets which are not Borel Sets. (See Exercise 20.12 for another approach.)
We can now see that our method of obtaining a collection of sets on which
m^* is a measure yields a vast collection of measurable sets.

Thus we have seen that there are many measurable sets which are
not Borel sets. Nevertheless, the following result shows that in some sense
every measurable set has a Borel set "almost" equal to it.

12.9 Theorem: Given a measurable set $A \subset E$, there exists a Borel set $B \subset E$
such that $m(A) = m(B)$. That is, B differs from A by a set of measure zero.

    Proof: Exercise 16.39.   □

In concluding section 12 we should point out that the complement
of the Cantor Set is open, dense in E (i.e. every open interval in E meets
it), and has measure 1. Also, it is possible to construct a closed nowhere
dense (i.e. containing no open interval of E) subset of E with measure
$> 1 - \epsilon$ where $\epsilon > 0$ is specified in advance (see Exercise 16.24).

13.** Necessary and Sufficient Conditions for a Set to be Measurable

There are several different ways to define the class M of measurable
sets in E. One of these is called the Carathéodory Criterion. In essence it
requires that for a set A to be measurable, outer measure must be additive
when A and $E \backslash A$ are intersected with any set X in E.

13.1 Theorem: (Carathéodory Criterion for Measurability) A set $A \subset E$ is
measurable if and only if $m^*(X) = m^*(X \cap A) + m^*(X \cap [E \backslash A])$ for
every subset $X \subset E$.

    Proof: Suppose A is measurable and $X \subset E$. Clearly $m^*(X)$
$\leqslant m^*(X \cap A) + m^*(X \cap [E \backslash A])$ by subadditivity of m^*.

To prove the reverse inequality, let $\epsilon > 0$ be given. Then there is
an open set $G \supset X$ such that $m(G) < m^*(X) + \epsilon$. Thus $m^*(X \cap A) +$

$m^*(X \cap [E\backslash A]) \leqslant m(G \cap A) + m(G \cap [E\backslash A]) = m(G) < m^*(X) + \epsilon$
since m is additive on disjoint measurable subsets of E. Since this inequality
holds for every $\epsilon > 0$, $m^*(X \cap A) + m^*(X \cap [E\backslash A]) \leqslant m^*(X)$.

The converse follows by letting $X = E$. □

One advantage of the Caratheodory Criterion is that it does not
require the concept of inner measure. Thus it can be used to define measure
and measurability for subsets of $\mathfrak{R}$.

Recall that in Lemma 10.1 we proved that if G_1 and G_2 are open in
E, then $m^*(G_1) + m^*(G_2) \geqslant m^*(G_1 \cup G_2) + m^*(G_1 \cap G_2)$. The reverse
inequality also holds (see Exercise 16.3) and we use this relationship to
prove the following useful criterion for a set to be measurable.

13.2 Theorem: A subset A of E is measurable if and only if for any $\epsilon > 0$,
there exist open sets G_1 and G_2 such that $G_1 \supset A$, $G_2 \supset E\backslash A$, and
$m(G_1 \cap G_2) < \epsilon$.

Proof: Let A be measurable and $\epsilon > 0$ be given. Then there exist
open G_1, G_2 in E with $G_1 \supset A$, $G_2 \supset E\backslash A$, and $m(G_1) < m^*(A) + \epsilon/2$,
and $m(G_2) < m^*(E\backslash A) + \epsilon/2$.

Now $m(G_1 \cap G_2) = m(G_1) + m(G_2) - m(G_1 \cup G_2) < m^*(A) + \epsilon/2$
$+ m^*(E\backslash A) + \epsilon/2 - m(G_1 \cup G_2)$. But $G_1 \cup G_2 = E$, so

$$m(G_1 \cap G_2) < m^*(A) + m^*(E\backslash A) - 1 + \epsilon.$$

Since A is measurable, $m^*(A) + m^*(E\backslash A) = 1$ and so $m(G_1 \cap G_2) < \epsilon$.
Conversely, given $\epsilon > 0$, suppose there exist open G_1, G_2 such that
$G_1 \supset A$, $G_2 \supset E\backslash A$, and $m(G_1 \cap G_2) < \epsilon$. Then

$$m^*(A) + m^*(E\backslash A) \leqslant m(G_1) + m(G_2) = m(G_1 \cup G_2)$$

$$+ m(G_1 \cap G_2) = 1 + \epsilon.$$

Thus $m^*(A) + m^*(E\backslash A) \leqslant 1$. Clearly $1 = m^*(E) \leqslant m^*(A) + m^*(E\backslash A)$ by
subadditivity, so equality holds and A is measurable. □

13.3 Corollary: A set $A \subseteq E$ is measurable if and only if for any $\epsilon > 0$, there
exist open $G \supset A$ and closed $F \subset A$ such that $m(G\backslash F) < \epsilon$.

Proof: Let $F = E\backslash G_2$ in the Theorem. □

There is also a standard criterion for measurability involving the symmetric difference of two sets. We include this in Exercise 16.27.

14. Lebesgue Measure for Bounded Sets

It would be easy (and we leave it to the reader) to repeat our development of Lebesgue measure for subsets of $[a,b]$, for each $a < b$. The definitions and proofs will be nearly identical to those for subsets of $E = [0,1]$. The most notable change is for inner measure:

$$m_*(A) = (b - a) - m^*([a,b] \backslash A), \qquad \text{for } A \subset [a,b].$$

So we will now assume that for each $a < b$, we have m^*, m_* defined on all subsets of $[a,b]$, we have the concept of measurability of a subset of $[a,b]$, and we have all the nice theorems developed in the last two chapters.

We still need to achieve all of these results for arbitrary bounded sets. For, if A is a bounded set, it is contained in infinitely many closed intervals. It might turn out that A would be measurable as a subset of $[a,b]$ and not measurable as a subset of $[c,d]$. Fortunately, this does not happen: $m^*(A)$, $m_*(A)$, and measurability of A do not change when we consider A as a subset of two different closed intervals.

First of all, if $G = \bigcup_i I_i$ is open and bounded, it is clear that the definition $m^*(G) = \sum_i m^*(I_i)$, where the I_i are pairwise disjoint open intervals, gives an unequivocal value, no matter what closed interval contains G. This is the basis of the following consistency result for the definition of $m^*(A)$.

14.1 Proposition: If $A \subset [a,b]$ and $A \subset [c,d]$, then $\mathrm{glb}\ \{m^*(G)|G$ open in $[a,b]$ and $A \subset G\} = \mathrm{glb}\ \{m^*(G)|G$ open in $[c,d]$ and $A \subset G\}$.

Proof: Both glb's are equal to $\mathrm{glb}\ \{m^*(G)|G$ open in $\mathcal{R}$ and $A \subset G\}$. See Exercise 16.28. □

The Proposition says that $m^*(A)$ is the same no matter what closed interval we take as containing it. The same is true of $m_*(A)$, as is shown in the next proposition.

14.2 Proposition: If $A \subset [a,b]$ and $A \subset [c,d]$, then

$$(b - a) - m^*([a,b] \setminus A) = (d - c) - m^*([c,d] \setminus A).$$

Proof: The result is easy if $A = \phi$. If $A \neq \phi$, then $A \subset [a,b] \cap [c,d]$ which is not empty. There are several ways this can happen—let us suppose that $a < c < b = d$. The remaining cases are left to the reader.

Let $\epsilon > 0$, and let G be open in $[c,d]$ with $([c,d] \setminus A) \subset G$ and $m^*(G) < m^*([c,d] \setminus A) + \epsilon$. Then $H = G \cup [a,c)$ is open in $[a,b]$, and $([a,b] \setminus A) \subset H$, so

$$m^*([a,b] \setminus A) \leqslant m^*(G \cup [a,c)) = m^*(G) + (c - a)$$

$$< m^*([c,d] \setminus A) + (c - a) + \epsilon.$$

Since ϵ was arbitrary, $m^*([a,b] \setminus A) \leqslant m^*([c,d] \setminus A) + (c - a)$.

On the other hand, if G' is open in $[a,b]$, $G' \supset ([a,b] \setminus A)$, and $m^*(G') < m^*([a,b] \setminus A) + \epsilon$, then $[a,c) \subset G'$, $H' = G' \setminus [a,c)$ is open in $[c,d]$, contains $[c,d] \setminus A$, and

$$m^*([c,d] \setminus A) \leqslant m^*(H') = m^*(G') - (c - a)$$

$$< m^*([a,b] \setminus A) - (c - a) + \epsilon. \qquad \square$$

From now on we will speak freely of measurability, m^*, and m_* for arbitrary bounded sets, and will feel free to use all the results we developed for subsets of $[0,1]$ in this more general context. Of particular importance, of course, is countable additivity which now has the following form.

14.3 Theorem: Let $\{A_i\}_{i=1}^{\infty}$ be a pairwise disjoint collection of measurable sets with $\bigcup_{i=1}^{\infty} A_i$ bounded. Then $\bigcup_{i=1}^{\infty} A_i$ is measurable and

$$m^*(\bigcup_{i=1}^{\infty} A_i) = \sum_{i=1}^{\infty} m^*(A_i).$$

In the remainder of the book we will use the numbers of theorems about measure on subsets of E to refer to analogous results about measure on bounded sets.

15.** Lebesgue Measure for Unbounded Sets

If A is unbounded, then we cannot apply our previous construction directly to define $m(A)$, since $m_*(A) = (b - a) - m^*([a,b]\backslash A)$ makes sense only if $A \subset [a,b]$. On the other hand, many unbounded sets intuitively should be measurable and have finite measure. For example, let $A = (0,\frac{1}{2}) \cup (1,1+\frac{1}{4}) \cup (2,2+\frac{1}{8}) \cup (3,3+\frac{1}{16}) \cup \cdots$. Intuitively we expect $m(A) = \frac{1}{2} + \frac{1}{4} + \frac{1}{8} + \frac{1}{16} + \cdots = 1$. How then should we define measurability and measure for unbounded sets?

The simplest solution is to truncate our unbounded set A by intersecting it with $[-n,n]$. This intersection will be bounded, and if measurable, $m(A \cap [-n,n])$ will be a finite number. Since $m(A \cap [-1,1]) \leqslant m(A \cap [-2,2]) \leqslant \cdots$, we are led to the definition

$$m(A) = \lim_{n \to \infty} m(A \cap [-n,n]).$$

This limit is either $+\infty$ or a finite non-negative real number.

15.1 Definition: A set A (not necessarily bounded) is measurable if $A \cap [-n,n]$ is measurable for every $n = 1,2, \cdots$. In that case

$$m(A) = \lim_{m \to \infty} m(A \cap [-n,n]).$$

Notice that this definition is consistent with our previous concept of measurability of A and the value of $m(A)$ for bounded sets A (Exercise 16.30). Notice also that $m(A) = \infty$ is a possibility. For example, $\mathcal{R}$ is measurable and $m(\mathcal{R}) = \infty$.

Our extended measure has the nice properties we are accustomed to, as exemplified by the following results.

15.2 Proposition: If $A \subset B$ are both measurable, then $m(A) \leqslant m(B)$.

Proof: Clearly $A \cap [-n,n] \subset B \cap [-n,n]$ for every n. $\qquad \square$

15.3 Theorem: (Countable Additivity) If $\{A_i\}_{i=1}^{\infty}$ is a collection of pairwise disjoint measurable sets, then $A = \bigcup_{i=1}^{\infty} A_i$ is measurable and

$$m(\bigcup_{i=1}^{\infty} A_i) = \sum_{i=1}^{\infty} m(A_i).$$

Proof: Clearly $A \cap [-n,n] = \bigcup_{i=1}^{\infty} (A_i \cap [-n,n])$. Each $A_i \cap [-n,n]$ is measurable, so $A \cap [-n,n]$ is measurable.

Now the formula $m(\bigcup_{i=1}^{\infty} A_i) = \sum_{i=1}^{\infty} m(A_i)$ is true if $m(A_i) = \infty$ for some i. Therefore, let $m(A_i) < \infty$ for all i. Then, since the sets

$$\{A_i \cap [-n,n]\}_{i=1}^{\infty}$$

are disjoint and measurable, countable additivity for bounded sets gives $m(A \cap [-n,n]) = \sum_{i=1}^{\infty} m(A_i \cap [-n,n]) \leqslant \sum_{i=1}^{\infty} m(A_i)$. Since this is true for every n; letting $n \to \infty$ we obtain $m(A) \leqslant \sum_{i=1}^{\infty} m(A_i)$.

On the other hand, if $\epsilon > 0$ and if k is given, then there is an n such that $m(A_i) < m(A_i \cap [-n,n]) + \epsilon/k$ for $i = 1, 2, \cdots, k$ (Why?). Therefore $\sum_{i=1}^{k} m(A_i) < \sum_{i=1}^{k} m(A_i \cap [-n,n]) + \epsilon \leqslant \sum_{i=1}^{\infty} m(A_i \cap [-n,n]) + \epsilon = m(A \cap [-n,n]) + \epsilon \leqslant m(A) + \epsilon$. Since ϵ and k were arbitrary, $\sum_{i=1}^{\infty} m(A_i) \leqslant m(A)$. $\qquad\square$

16. Exercises

16.1 Prove Lemma 10.1 for G_1, G_2 finite unions of disjoint open intervals. (Hint: see Exercise 5.21. Verify that $\int_0^1 \chi_{G_1}(x)\,dx = m^*(G_1)$, where this is the ordinary Riemann integral.)

16.2 Prove Lemma 10.2(2).

16.3 Prove that if A_1 and A_2 are measurable, then

$$m^*(A_1) + m^*(A_2) = m^*(A_1 \cup A_2) + m^*(A_1 \cap A_2).$$

(Hint: 10.2, 10.3.)

16.4 Using Corollary 10.3 and induction, prove that finite unions and intersections of measurable sets are measurable.

16.5 (a) Using Corollary 10.3, show that if A and B are measurable subsets of E, then $A \setminus B$ is measurable.

(b) Use Countable Additivity (10.6) to show that if $B \subset A$ and B and A are measurable subsets of E, then $m^*(A \setminus B) = m^*(A) - m^*(B)$.

16.6 Show that for any measurable set $A \subset E$ with $m(A) > 0$, there are real numbers x and $y \in A$ such that $x - y$ is rational. (Hint: see the construction in Example 7.7.)

16.7 Prove Corollary 10.10.

16.8 If $m(A) = 0$ and $f{:}E \to E$ has bounded derivative, show that $m(f(A)) = 0$. (Hint: use the mean value theorem on each open interval of an open set G containing A.)

16.9 If $S \subset E$, then there is a measurable set $A \subset E$ such that $S \subset A$ and $m(A) = m^*(S)$. (Hint: take an intersection of a decreasing sequence of approximating open sets.)

16.10 Prove Proposition 12.2.

16.11 Verify that the following collections are countably additive classes of sets:

(a) the collection of all subsets of E.
(b) the collection of all measurable subsets of E.

16.12 Part (3) of Definition 12.1 is called "closure under countable unions." Show that the class of measurable subsets of E is *not* closed under *un*countable unions; i.e., there exists an uncountable collection of measurable subsets of E whose union is not measurable.

16.13 Show that any arbitrary intersection of countably additive classes is a countably additive class.

16.14 (a) Show that every open set of E is a Borel set.
(b) Show that every closed set of E is a Borel set.
(c) Show that every half-open interval $(a,b] \subset E$ is a Borel set.
(d) Show that $Q \cap E$ is a Borel set.

16.15 Show that if $f{:}E \to E$ is continuous and $A \subset E$ is a Borel set, then $f^{-1}(A) = \{x \in E | f(x) \in A\}$ is a Borel set. (Hint: show that the collection of all sets A such that $f^{-1}(A)$ is a Borel set, is a countably additive class which contains all the open intervals.)

16.16 Let a be an integer bigger than 1. An *a-ary expression* for a number $x \in E$ is an expression of the form . $a_1 a_2 a_3 a_4 \ldots$, where each $0 \leqslant a_i < a$, and where $x = \dfrac{a_1}{a} + \dfrac{a_2}{a^2} + \dfrac{a_3}{a^3} + \ldots$. The series on the right converges by comparison with the convergent geometric series $\sum\limits_{n=1}^{\infty} \dfrac{a}{a^n}$. The decimal $(a = 10)$, binary $(a = 2)$, and ternary $(a = 3)$ expressions for a number in E are special cases.

(a) Find the rational number whose 4-ary expression is .2222. . . . .
(b) Find the rational number whose 5-ary expression is .2120202020. . . .
(c) Find the binary expression for $\frac{3}{4}$.
(d) Find the ternary expression for $\frac{1}{4}$.
(e) Show that every number $x \in E$ has an a-ary expression. (Hint: let

$$a_1 = \text{the greatest integer } b \text{ such that } \frac{b}{a} \leqslant x.$$

Then

$$a_2 = \text{the greatest integer } b \text{ such that } \frac{a_1}{a} + \frac{b}{a^2} \leqslant x, \text{ etc.}$$

Prove that the resulting series converges to x.)

(f)   Show that for any a-ary expression of the form $. a_1 a_2 \ldots a_n 000 \ldots ,$ with $a_n \neq 0$, there is a different expression with the same value. (In ternary, for example, $.02000 \ldots = .01222222 \ldots .$)

16.17 Let $B = \{x \in E \,|\, \text{a binary expression for } x \text{ has a 0 in every even position}\}$. Show that B is measurable and $m(B) = 0$. (Hint: write B as the intersection of a decreasing sequence of sets.)

16.18 Let $f : C \to E$ be defined by $f(. a_1 a_2 a_3 \ldots) = \frac{a_1}{2} \frac{a_2}{2} \frac{a_3}{2} \ldots$, where the expression on the left is ternary with no 1's, and the expression on the right is binary. Prove that f is onto but not $1 - 1$. (See Exercise 16.16(f).)

16.19 Prove that the function f in Exercise 16.18 is continuous.

16.20 Prove that C is uncountable. (Hint: use the ternary characterization of C, and a modification of the usual diagonal argument used to show that $(0,1)$ is uncountable.)

16.21 (a)   Prove that the Cantor set C is *nowhere dense in E* (contains no interval of E.).
  (b)   Prove that every point of C is a limit point of C; that is, for every $a \in C$ and every $\epsilon > 0$, there is $b \in C$ with $|a - b| < \epsilon$. A closed set with this property is said to be *perfect*.

16.22 Show that if A is nowhere dense in E (Exercise 16.21), then $E \backslash A$ is dense in E (i.e., every point of A is a limit point of $E \backslash A$).

16.23 Construct a perfect nowhere dense set $D \subset E$ such that $m(D) = \frac{1}{2}$. (Hint: follow the construction of Example 12.5, but throw away less than the middle third at each step.)

16.24 Show that for any $\epsilon > 0$, there is a perfect nowhere dense set $C_\epsilon \subset E$ such that $m(C_\epsilon) > 1 - \epsilon$.

16.25 (a)   Prove that $A_1 \cup A_2$ is measurable if A_1 and A_2 are measurable, using only the Caratheodory criterion (Theorem 13.1) and properties of m^*.
  (b)   Repeat part (a), using the criterion of Theorem 13.2.

16.26 Prove that for any $A \subset E$, $m_*(A) = \text{lub } \{m^*(F) | F \text{ is closed and } F \subset A\}$.

16.27 Given $A, B \subset E$, define the *symmetric difference* $A \triangle B$, by

$$A \triangle B = (A \backslash B) \cup (B \backslash A).$$

(a) Prove that if A and B are measurable, so is $A \triangle B$.

(b) Prove that A is measurable if and only if for every $\epsilon > 0$, there is a finite union $\overset{n}{\underset{i=1}{\cup}} I_i$ of open intervals such that $m^*(A \triangle \overset{n}{\underset{i=1}{\cup}} I_i) < \epsilon$.

(c) Prove that if A is measurable and $m^*(A \triangle B) = 0$, then B is measurable and $m(A) = m(B)$.

16.28 Prove that for $A \subset [a,b]$,

$$\text{glb } \{m^*(G) | G \text{ open in } [a,b] \text{ and } A \subset G\}$$

$$= \text{glb } \{m^*(G) | G \text{ open in } \mathcal{R} \text{ and } A \subset G\}.$$

16.29 Show that if A is a measurable set with positive measure, then A has a non-measurable subset. (Assume A is bounded if you need to. Follow the construction of Example 7.7. (See also Corollary 10.8.) You may wish to try $A = [a,b]$ first.)

16.30 Show that if A is bounded, then A is measurable if and only if $A \cap [-n,n]$ is measurable for every positive interger n, and in that case

$$m(A) = \lim_{n \to \infty} m(A \cap [-n,n]).$$

16.31 Find $m(Q)$, $m(\mathcal{R} \backslash Q)$, $m(Z)$ (where Z is the set of integers).

16.32 Prove that (even for unbounded sets) if A and B are measurable, so are $A \cup B$, $A \cap B$, and $A \backslash B$.

16.33 Prove that if A and B are measurable (not necessarily bounded), then

$$m(A) + m(B) = m(A \cup B) + m(A \cap B).$$

16.34 Prove that if A_i is measurable for $i = 1, 2, \ldots$, then $\overset{\infty}{\underset{i=1}{\cup}} A_i$ and $\overset{\infty}{\underset{i=1}{\cap}} A_i$ are measurable, and

$$m(\overset{\infty}{\underset{i=1}{\cup}} A_i) \leq \overset{\infty}{\underset{i=1}{\Sigma}} m(A_i).$$

16.35 Prove Corollary 10.9 in the case of unbounded sets A_i, $i = 1, 2, \ldots$.

16.36 Show that Corollary 10.10 is *false* for unbounded sets A_i, $i = 1, 2, \ldots$. Where does the proof of Corollary 10.10 break down in this case?

16.37 Show that every compact set is measurable with finite measure.

16.38 Show that if A is bounded, $m^*(cA) = |c| m^*(A)$, where $cA = \{cx \mid x \in A\}$. Also show that cA is measurable if A is measurable.

16.39 Prove Theorem 12.9.

CHAPTER 4

Measurable Functions

17. Definition of Measurable Function

At the beginning of Chapter 2, we indicated that our definition of the Lebesgue integral for $f:[a,b] \to \Re$ would require finding the measure of sets of the form $E_i = \{x \in [a,b] \mid y_{i-1} \leqslant f(x) < y_i\}$. We have nearly accomplished this, but unfortunately we found that not all sets can successfully be assigned a measure. We have shown that if we restrict ourselves to *measurable* sets, then we do have a measure m. Therefore, we must require that sets of the form E_i be measurable. Since E_i must be measurable for arbitrary y_{i-1}, y_i, we evidently must restrict ourselves to functions f for which $\{x \in [a,b] \mid c \leqslant f < d\}$ is measurable for every $c < d$. It turns out that we can relax this condition slightly and still get the same class of functions (see Proposition 17.2).

We would like to allow our functions more general types of domains than intervals of the form $[a,b]$, so with section 14 of Chapter 3 in mind we make the following convention.

From now on A will always be a bounded measurable set.

17.1 Definition: Let A be a bounded measurable subset of $\Re$. Then $f:A \to \Re$ is measurable on A if $\{x \in A \mid c < f(x)\}$ is measurable for every real number c.

Notice that in the definition there are infinitely many sets to check for measurability, one for each real number c. We will delete the phrase "on A" if it is clear from context what the set A is. Notice also that the function f may be unbounded on A.

Now we need to show that the class of measurable functions satisfies the requirements we discussed before the definition.

17.2 Proposition: Let A be bounded and measurable and $f:A \to \Re$. Then the following are equivalent:

(1) $\{x \in A \mid c < f(x)\}$ is measurable for all c (f is measurable);
(2) $\{x \in A \mid c \leq f(x)\}$ is measurable for all c;
(3) $\{x \in A \mid f(x) < c\}$ is measurable for all c;
(4) $\{x \in A \mid f(x) \leq c\}$ is measurable for all c;
(5) $\{x \in A \mid c < f(x) \leq d\}$ is measurable for all $c < d$;
(6) $\{x \in A \mid c \leq f(x) < d\}$ is measurable for all $c < d$;
(7) $\{x \in A \mid c \leq f(x) \leq d\}$ is measurable for all $c < d$;
(8) $\{x \in A \mid c < f(x) < d\}$ is measurable for all $c < d$.

(In other words, any of the conditions (1) – (8) could be taken as the definition of measurability. After proving their equivalence, we will use the conditions interchangeably.)

Proof: (1) $\Rightarrow$ (2). Surely $c \leq f(x)$ if and only if for every natural number n, $c - 1/n < f(x)$. Therefore

$$\{x \in A \mid c \leq f(x)\} = \{x \in A \mid (\forall n)(c - 1/n < f(x))\}$$

$$= \bigcap_{n=1}^{\infty} \{x \in A \mid c - 1/n < f(x)\},$$

but the latter set is measurable by (1) and preservation of measurability under countable intersections (Corollary 10.11 generalized to arbitrary bounded sets).

(2) $\Rightarrow$ (3). It is clear that $\{x \in A \mid f(x) < c\} = A \setminus \{x \in A \mid c \leq f(x)\}$, and the latter set is measurable by (2) and preservation of measurability under relative complements. (Corollary 8.4 and Corollary 10.3).

(3) $\Rightarrow$ (4) Exercise 20.1.

(4) $\Rightarrow$ (1) Exercise 20.2.

We may now use (1) - (4) interchangeably in the remainder of the proof.

(1) - (4) $\Rightarrow$ (5). Notice that $\{x \in A \mid c < f(x) \leq d\} = \{x \in A \mid c < f(x)\}$ $\cap \{x \in A \mid f(x) \leq d\}$, both of which are measurable by (1) and (4) respectively. The result follows from preservation of measurability under intersections (Corollary 10.3).

(5) $\Rightarrow$ (1). Use the fact that

$$\{x \in A \mid c < f(x)\} = \bigcup_{n=1}^{\infty} \{x \in A \mid c < f(x) \leq c + n\}.$$

The remainder of the proof is left to the reader. $\qquad\square$

In contrast to the Proposition, if f is a measurable function, then $\{x \in A \mid c = f(x)\}$ is measurable for each real number c, but this condition is not sufficient for measurability (see Exercise 20.4).

17.3 Example: This function χ_B is measurable on $A \supset B$ if and only if B is measurable. In fact,

$$\{x \in A \mid \chi_B(x) > c\} = \begin{cases} \phi & \text{if } c \geq 1. \\ B & \text{if } 0 \leq c < 1. \\ A & \text{if } c < 0. \end{cases}$$

This means that χ_Q and $\chi_{A \setminus Q}$ are measurable functions on A but χ_V is not measurable on $[0,1]$, where V is the non-measurable set in Example 7.7.

17.4 Example: For every set A with $m(A) > 0$, there is a function $f : A \to \Re$ which is not measurable on A (see Exercise 20.4).

Now we mention a characterization of measurability which is reminiscent of continuity. Recall that a function $g : A \to \Re$ is continuous on A if and only if $g^{-1}(G)$ is open in A for every open set $G \subset \Re$. If we relax this condition to require only that $g^{-1}(G)$ be measurable, then we have measurability of g.

17.5 Theorem: A function $f : A \to \Re$ is measurable if and only if $f^{-1}(G)$ is measurable for every open set $G \subset \Re$.

Proof: Let f be measurable. Let $G \subset \mathcal{R}$ be open. If $G = \mathcal{R}$, then $A = f^{-1}(G)$ is measurable. If $G \neq \mathcal{R}$, then Theorem 7.2 yields $G = \bigcup_i I_i$, where the I_i are disjoint open intervals, at most two of which are infinite of the form $(-\infty, c)$ or (d, ∞) for c, d real numbers. Since the order of the I_i's is irrelevant, let us suppose that $I_1 = (-\infty, c)$ and $I_2 = (d, \infty)$ and all the remaining I_i's are of the form (a_i, b_i) for real numbers a_i and b_i.

Then $f^{-1}(G) = \bigcup_{i=1}^{\infty} f^{-1}(I_i)$, and by the preservation of measurability under unions (Corollary 10.11), it is sufficient to show that $f^{-1}(I_i)$ is measurable for each $i = 1, 2, 3, \cdots$. But $f^{-1}((-\infty, c)) = \{x \in A \mid f(x) < c\}$; $f^{-1}((d, \infty)) = \{x \in A \mid f(x) > d\}$; and $f^{-1}((a_i, b_i)) = \{x \in A \mid a_i < f(x) < b_i\}$ are all measurable since f is measurable.

For the converse see Exercise 20.11. □

Another possible analogy with continuity would be to expect that $f^{-1}(B)$ should be measurable for every measurable set B. However there is a measurable function f for which this is not true. (See Exercise 20.12.)

Theorem 17.5 has a couple of immediate corollaries.

17.6 Corollary: If $f:A \to \mathcal{R}$ is continuous, then f is measurable.

17.7 Corollary: If $f:A \to \mathcal{R}$ is measurable and $g:f(A) \to \mathcal{R}$ is continuous, then $g \circ f$ is measurable.

Proof: Exercise 20.13. □

Surprisingly, composition in the opposite order does not necessarily yield measurability. There is a measurable function g and a continuous function f such that $g \circ f$ is not measurable. (See Exercise 20.12.) It follows that the composition of two measurable functions need not be measurable.

18. Preservation of Measurability for Functions

Measure theory provides a concept of "almost equality" for functions which is very useful.

18.1 Definition: We will say that f equals g *almost everywhere* (abbreviated a.e.) on A if the set where they differ has measure zero. That is, $f = g$ (a.e.)

on A if $m(\{x \in A \mid f(x) \neq g(x)\}) = 0$.

Because subsets of sets of measure zero also have measure zero, this is equivalent to saying that there is a set $B \subset A$ with $m(B) = 0$ and $f(x) = g(x)$ for every $x \in A \backslash B$.

We will extend this terminology to other circumstances as well. In general "almost everywhere" will mean "except on a set of measure zero." Thus, "f is continuous a.e. on A" means "there is a subset $B \subset A$ such that $m(B) = 0$ and f is continuous on $A \backslash B$."

Now a simple but sometimes useful result.

18.2 Proposition: If f is measurable on A and $f = g$ (a.e.) on A, then g is measurable on A.

    Proof: Let $B = \{x \in A \mid f(x) \neq g(x)\}$. Then $m(B) = 0$ and

$$\{x \in A \mid g(x) > c\} = \{x \in B \mid g(x) > c\} \cup \{x \in A \backslash B \mid g(x) > c\}$$

$$= \{x \in B \mid g(x) > c\} \cup \{x \in A \backslash B \mid f(x) > c\}.$$

In this last union, the first set is a subset of B and hence is measurable (with measure 0). The second set is $\{x \in A \mid f(x) > c\} \cap (A \backslash B)$, hence is measurable.   □

Measurability is also preserved under many common manipulations of functions.

18.3 Proposition: If f is measurable on A, so are $|f|$, f^2, $1/f$ (if $f(x) \neq 0$ for all $x \in A$), and $\sqrt{f}$ (if $f(x) \geq 0$ for all $x \in A$).

    Proof: This follows immediately from Corollary 17.7. For example $|f| = g \cdot f$, where $g(x) = |x|$, a continuous function.   □

You should be able to extend the list in the above Proposition ad infinitum (see Exercise 20.14).

The same nice behavior also holds under algebraic combination of functions.

18.4 Proposition: If f and g are measurable on A, so are $f + g$, fg, and f/g (for g non-zero).

Proof: A slick proof like that of the previous proposition is possible, but we would have to consider measure on sets in $\mathcal{R}^2$. Directly, we note that for fixed $x \in A$,

$$f(x) + g(x) > c \text{ if and only if } f(x) > c - g(x) \text{ if and only if}$$

there is a rational number q such that $f(x) > q > c - g(x)$.

Therefore,

$$\{x \in A \,|\, f(x) + g(x) > c\} = \bigcup_{q \in Q} \{x \in A \,|\, f(x) > q \text{ and } q > c - g(x)\}$$

$$= \bigcup_{q \in Q} (\{x \in A \,|\, f(x) > q\} \cap \{x \in A \,|\, g(x) > c - q\}).$$

All the sets involved are measurable since f and g are measurable and Q is countable.

For fg, notice that $fg = \frac{1}{4}(f + g)^2 - \frac{1}{4}(f - g)^2$. The result follows from the preceding. (How do you deal with the $\frac{1}{4}$?)

Finally, $f/g = f(1/g)$. $\square$

18.5 Corollary: If f and g are measurable, then $\{x \in A \,|\, f(x) > g(x)\}$ is measurable.

Proof: Exercise 20.21. $\square$

Most important for us will be preservation of measurability under limits. The following results, in fact, form the underpinnings of the basic convergence theorems involving the Lebesgue Integral (see Chapter 6).

18.6 Theorem: If $f_n : A \to \mathcal{R}$ is measurable for each $n = 1, 2, \cdots$ and if f is the pointwise limit of $\{f_n\}$ (that is, $f(x) = \lim_{n \to \infty} f_n(x)$ for all $x \in A$), then f is measurable on A.

Proof: We must carefully analyze the meaning of the limit. Let $x \in A$. Then, if $f(x) > a$, let p be a natural number such that $f(x) > a + 1/p$. Then for large enough n, $f_n(x) > a + 1/p$. That is,

$$(\exists p)(\exists N)(\forall n > N)[f_n(x) > a + 1/p].$$

Conversely, if this condition is satisfied, then $f(x) = \lim_{n \to \infty} f_n(x) \geq a + 1/p > a$.

Therefore,

$$\{x \in A \,|\, f(x) > a\} = \bigcup_{p=1}^{\infty} \bigcup_{N=1}^{\infty} \bigcap_{n=N+1}^{\infty} \{x \in A \,|\, f_n(x) > a + 1/p\},$$

and this set is measurable. □

Related to this result is the following.

18.7 Proposition: If $\{f_n\}$ is a pointwise bounded sequence of functions (that is, $\{f_n(x) \,|\, n = 1,2, \cdots \}$ is bounded for each $x \in A$) each measurable on A, then $f(x) = \text{lub } \{f_n(x) \,|\, n = 1,2, \cdots \}$ is measurable on A. Similarly, $glb \{f_n(x) \,|\, n = 1,2, \cdots \}$ is measurable on A.

Proof: This follows from the equation

$$\{x \,|\, f(x) > a\} = \bigcup_{n=1}^{\infty} \{x \,|\, f_n(x) > a\}. \qquad \square$$

18.8 Corollary: Under the conditions of the proposition, $g(x) = \overline{\lim_{n \to \infty}} f_n(x)$ and $h(x) = \underline{\lim_{n \to \infty}} f_n(x)$ are measurable.

Proof: Use the relationship $g(x) = \lim_{k \to \infty} (\text{lub } \{f_n(x) \,|\, n \geq k\})$, and a similar relationship for $h(x)$. □

19. Simple Functions

In discussing the Lebesgue Integral in the next chapter, we will have need of some particularly uncomplicated measurable functions. These will play a role analogous to that of step functions in the Riemann theory.

19.1 Definition: A *simple function* $g : A \to \Re$ is a measurable function with finitely many values.

19.2 Proposition: A function $g : A \to \Re$ is simple if and only if there are real numbers $b_1, \cdots, b_n$ and pairwise disjoint measurable sets $B_1, \ldots, B_n$ with $A = \bigcup_{i=1}^{n} B_1$ and $g = \sum_{i=1}^{n} b_i \chi_{B_i}$.

Proof: If g is simple on A, let its distinct values be $b_1, \cdots, b_n$. Then by Exercise 20.4, $B_i = \{x \in A \,|\, f(x) = b_i\}$ is measurable for each $i = 1, 2, \cdots, n$. Furthermore, $B_i \cap B_j = \phi$ for $i \neq j$ and $\bigcup_{i=1}^{n} B_i = A$. Therefore, $g = \sum_{i=1}^{n} b_i \chi_{B_i}$, since if $x \in B_j, \chi_{B_i}(x) = \{ \begin{smallmatrix} 1 & \text{if } i = j \\ 0 & \text{if } i \neq j \end{smallmatrix}$, so that $g(x) = b_j = \sum_{i=1}^{n} b_i \chi_{B_i}(x)$.

For the converse, see Exercise 20.27. $\qquad\square$

If g is simple, then any sum of the form in the Proposition will be called a *representation of g*. Clearly every simple function can have many different representations. (How many?)

Simplicity is preserved by algebraic combinations.

19.3 Proposition: If $f{:}A \to \mathcal{R}$, $g{:}A \to \mathcal{R}$ are simple, so are $f + g$, fg, and f/g (for g non-zero).

Proof: Exercise 20.30. $\qquad\square$

Simplicity is *not* preserved under limits. However, the pointwise limit of a sequence of simple functions is measurable by Theorem 18.6. The converse is also true.

19.4 Theorem: A function $f{:}A \to \mathcal{R}$ is measurable if and only if f is the pointwise limit of a sequence of simple functions on A.

Proof: ($\Leftarrow$). Theorem 18.6.

($\Rightarrow$). Let P_n be a partition of $[-n, n]$ (on the y-axis) obtained by taking equal sub-intervals of length $1/n$. (Thus

$$P_n = (y_0, \cdots, y_{2n^2}),$$

where $y_i = -n + i/n$ for $i = 1, 2, \cdots, 2n^2$.)

Now let $B_i = \{x \in A \,|\, y_{i-1} \leqslant f(x) < y_i\}$, and let $f_n = \sum_{i=1}^{2n^2} y_{i-1} \chi_{B_i}$. Then f_n is simple.

Now given $x_0 \in A$, for each n big enough so that $f(x_0) \in [-n, n]$, $y_{i-1} \leqslant f(x_0) < y_i$ for some $i = 1, 2, \cdots, 2n^2$. Thus $x_0 \in B_i$ and $|f(x_0) - f_n(x_0)| = |f(x_0) - y_{i-1}| < |y_i - y_{i-1}| = 1/n$.

So $\lim_{n \to \infty} f_n(x_0) = f(x_0)$ for all $x_0 \in A$. ☐

If f is *bounded below* (above), then it is possible to modify the construction in the proof of the theorem so that the sequence f_n is increasing (decreasing). See Exercise 20.32. If f is bounded above *and* below, we can do even better.

19.5 Corollary: A function $f:A \to \mathcal{R}$ is bounded and measurable if and only if f is the *uniform* limit of a sequence of simple functions.

Proof: If f is bounded, then for large enough n, $f(x) \in [-n,n]$ for all $x \in A$. Therefore, the calculation in the proof of the Theorem shows that $|f_n(x) - f(x)| < 1/n$ for all $x \in A$. Uniformity follows.

For the converse, see Exercise 20.35. ☐

20. Exercises

20.1 Prove $(3) \Rightarrow (4)$ in Proposition 17.2.

20.2 Prove $(4) \Rightarrow (1)$ in Proposition 17.2.

20.3 Prove the remainder of Proposition 17.2.

20.4 (a) Prove that if f is measurable, then $\{x \in A \mid c = f(x)\}$ is measurable for each real number c.
  (b) Show that the converse of (a) is false if $m(A) > 0$. (Hint: by Exercise 16.29, there is a non-measurable set $B \subset A$.)

20.5 Prove that if $m(A) = 0$, then every function is measurable on A.

20.6 If B is any set, $f:B \to \mathcal{R}$, and $\{x \in B \mid f(x) < c\}$ is measurable for every real number c, show that B is measurable.

20.7 Prove directly from Proposition 17.2 that each of the following functions is measurable on $[0,1]$.
  (a) $f(x) = 3$ for all $x \in [0,1]$.
  (b) $f(x) = x$ for all $x \in [0,1]$.
  (c) $f(x) = \begin{cases} 1/x \text{ if } x \in (0,1] \\ 0 \text{ if } x = 0. \end{cases}$

20.8 If $B \subset A$, B measurable, $f:A \to \mathcal{R}$ measurable on A, then f is measurable on B.

20.9 If $f:A \cup B \to \mathcal{R}$ is measurable on A and on B, prove that f is measurable on $A \cup B$, on $A \cap B$, and on $A \backslash B$.

20.10 If f is measurable on A and $m(B) = 0$, show that f is measurable on $A \cup B$.

20.11 Prove the remainder of Theorem 17.5; if $f:A \to \mathcal{R}$ and $f^{-1}(G)$ is measurable for every open set $G \subset \mathcal{R}$, then f is measurable on A.

20.12 Let C be the Cantor set (Example 12.5 and Exercise 16.21). Let $D \subset [0,1]$ be a nowhere dense measurable set with $m(D) > 0$ (Exercise 16.22). Then there is a non-measurable set $B \subset D$ (Exercise 16.29). At each stage of the construction of C and of D, a certain finite number of open intervals of $[0,1]$ are deleted (put into $[0,1] \backslash C$ or $[0,1] \backslash D$). Let g map the intervals put into $[0,1] \backslash D$ at the nth stage linearly onto the intervals put into $[0,1] \backslash C$ at the nth stage, for $n = 1,2,\ldots$. (See illustration.) Thus g is monotone and defined at every element of $[0,1] \backslash D$, mapping onto $[0,1] \backslash C$.

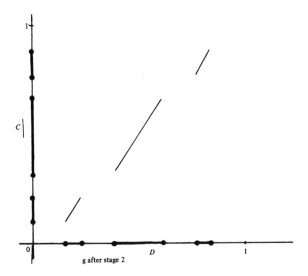

g after stage 2

(a) Using the fact that g is increasing and that $[0,1] \backslash D$ is dense in $[0,1]$ (Exercise 16.22), prove that for every $x_0 \in D$,

$$\lim_{x \to x_0^-} g(x) = \text{lub } \{g(x) | x \in [0,1] \backslash D \text{ and } x < x_0\}$$

and

$$\lim_{x \to x_0^+} g(x) = \text{glb } \{g(x) | x \in [0,1] \backslash D \text{ and } x > x_0\}.$$

(b) Use (a) to show that for $x_0 \in D$, $\lim\limits_{x \to x_0^-} g(x) \leqslant \lim\limits_{x \to x_0^+} g(x)$.

(c) Show that if $a = \lim\limits_{x \to x_0^-} g(x) < \lim\limits_{x \to x_0^+} g(x) = b$ for some $x_0 \in D$, then C would contain the interval (a,b), contradicting the fact that C is nowhere dense. Therefore $\lim\limits_{x \to x_0^-} g(x) = \lim\limits_{x \to x_0^+} g(x) = \lim\limits_{x \to x_0} g(x)$, and the function

$$f(x) = \begin{cases} g(x) \text{ if } x \in [0,1] \backslash D \\ \lim_{x' \to x} g(x') \text{ if } x \in D \end{cases}$$

is continuous and monotone.

(d) Let $B_0 = f(B)$. Show that B_0 is measurable but $f^{-1}(B_0) = B$ is not, even though f is a measurable function.

Since $f^{-1}(B_0)$ is not measurable, it is not a Borel set. By Exercise 16.15, B_0 is not a Borel set, even though it is measurable.

(e) Using the fact that $\chi_B = \chi_{B_0} \cdot f$, show that composition in the opposite order in Corollary 17.7 is false.

20.13 Prove Corollary 17.7.

20.14 Show that if f is measurable, so is the function g defined by $g(x) = e^{f(x)}$.

20.15 Show that if f is continuous a.e. on a compact set K, then for every $\epsilon > 0$, there is a set $A \subset K$ such that f is bounded on A, and $m(K \backslash A) < \epsilon$. (Hint: modify the usual compactness argument for continuous functions.)

20.16 (a) Prove or disprove: if $f = g$ a.e., and g is continuous, then f is continuous a.e. .

(b) Prove or disprove the converse of (a).

20.17 Show that if f is continuous a.e. on a bounded measurable set A, then f is measurable. (By Exercise 20.16(b), you may not use Proposition 18.2.)

20.18 Show that "equality a.e." is an equivalence relation on the class of all functions on a bounded measurable set A.

20.19 Show directly from Proposition 17.2 that $|f|, f^2, 1/f (f \neq 0)$ are measurable if f is measurable.

20.20 Prove or disprove: if $|f|$ is measurable, so is f.

20.21 Prove Corollary 18.5.

20.22 Show that f is measurable on A if and only if f^2 is measurable and $\{x \in A | f(x) > 0\}$ is measurable.

20.23 If f and g are measurable on A, show that $\{x \in A | f(x) = g(x)\}$ is measurable.

20.24 Let f and g be measurable on A, and define a function h by

$$h(x) = \begin{cases} \dfrac{f(x) g(x)}{f(x) + g(x)} & \text{if } f(x) + g(x) \neq 0 \\ 0 & \text{if } f(x) + g(x) = 0. \end{cases}$$

Show that h is measurable on A.

20.25 Let f_n be measurable on bounded measurable set A, for $n = 1,2,\ldots$. If $B = \{x \in A \,|\, \{f_n(x)\}$ converges$\}$ show that B is measurable. (Hint: look at the Cauchy criterion for convergence.)

20.26 Let f_n be measurable on A for $n = 1,2,\ldots$. Let $f_n \to f$ pointwise a.e. on A. Show that for every real number $c > 0$,

$$\lim_{n \to \infty} m(\{x \in A \,|\, |f(x) - f_n(x)| \geqslant c\}) = 0.$$

(Hint: Note that $m(\bigcap_{k=1}^{\infty} \bigcup_{n=k}^{\infty} \{x \in A \,|\, |f(x) - f_n(x)| \geqslant c\}) = 0$.)

20.27 Prove the remainder of Proposition 19.2; if $g = \sum_{i=1}^{n} b_i \chi_{B_i}$, where the B_i are pairwise disjoint measurable sets with union A, then g is simple on A.

20.28 Show that every step function on $[a,b]$ is simple, but the converse is false.

20.29 Give an example of a simple function and two different representations of it.

20.30 Prove Proposition 19.3. Given representations $f = \sum_{i=1}^{m} b_i \chi_{B_i}$ and $g = \sum_{k=1}^{n} c_k \chi_{C_k}$, find explicit representations of $f + g$, fg, $f/g(g \neq 0)$.

20.31 Suppose g is simple and f is any function. Under what conditions is $f \circ g$ simple? When is $g \circ f$ simple?

20.32 Prove that if f is measurable on A and f is bounded below, then f is the pointwise limit of an *increasing* sequence of simple functions. (Hint: at stage n, partition $[-n,n]$ into $2n \cdot 2^n$ equal subintervals. Thus, in going from stage n to stage $n + 1$, each subinterval is split in half.)

20.33 Find an increasing sequence of simple functions on $[0,1]$ with limit $f(x) = x$.

20.34 (a) Show that the simple function χ_C (where C is the Cantor set) is the pointwise limit of a *decreasing* sequence of step functions, but not of an *increasing* sequence of step functions. (Hint: for the second part, you need the fact that C is uncountable and nowhere dense (Exercise 16.21).)

   (b) Show how to alter χ_C to create a simple function which is not the pointwise limit of any monotone sequence of step functions.

20.35 Prove the remainder of Corollary 19.5; if f is the *uniform* limit of simple functions, then f is bounded and measurable.

20.36 Prove that if f is monotone on a bounded measurable set A, then f is measurable on A.

CHAPTER 5

The Lebesgue Integral

21. The Lebesgue Integral for Bounded Measurable Functions

Let f be a bounded measurable function on a bounded measurable set $A \subset \mathfrak{R}$. We wish to define the Lebesgue integral of f on A. As we said in Chpater 2, this involves partitioning the range of f, rather than the domain as for the Riemann integral. More precisely, we need to partition an interval $[\ell, u]$ such that $\ell \leqslant f(x) < u$ for all $x \in A$. This may be accomplished by taking

$$\ell = \mathrm{glb} \, \{f(x) | x \in A\},$$

and

$$u = 1 + \mathrm{lub}\{f(x) | x \in A\}.$$

The evident arbitrariness of this choice will be eliminated as we proceed; that is, we will see that any choice of ℓ and u—as long as it satisfies the inequalities—will yield the same integral.

Now for our approximation.

21.1 Definition: If f is bounded measurable function on a bounded measurable set $A \subset \mathfrak{R}$, if $P = (y_0, y_1, \ldots, y_n)$ is a partition of

$$[\ell, u] = [\mathrm{glb}\{f(x) | x \in A\}, \, 1 + \mathrm{lub}\{f(x) | x \in A\}],$$

and if

$$y_i^* \in [y_{i-1}, y_i] \text{ for } i = 1, \ldots, n,$$

then we will call

$$L(f,P) = \sum_{i=1}^{n} y_i^* \cdot m(\{x \in A \,|\, y_{i-1} \leqslant f(x) < y_i\})$$

a *Lebesgue sum* of f relative to P.

As indicated in Chapter 2, $L(f,P)$ is an approximation to the area under the graph of f (if f is non-negative). It is the sum of areas of pseudo-rectangles whose bases are not necessarily intervals, but rather are measurable sets of the form $A_i = \{x \in A \,|\, y_{i-1} \leqslant f(x) < y_i\}$. Note that the inequalities $y_{i-1} \leqslant f(x) < y_i$ are chosen so that the sets A_i are pairwise disjoint and $A = A_1 \cup A_2 \cup \ldots \cup A_n$. Now we take the limit of $L(f,P)$ as $\|P\| \to 0$.

21.2 Definition: A bounded measurable function $f:A \to \Re$ is *Lebesgue integrable on A* if there is a real number L such that for every $\epsilon > 0$, there exists a $\delta > 0$ such that if $\|P\| < \delta$, and if $L(f,P)$ is a Lebesgue sum of f relative to P, then $|L(f,P) - L| < \epsilon$.

The reader should compare this with the definition of Riemann integrability (Definition 1.3). As for the Riemann case, we have uniqueness: at most one number L can satisfy the definition. This unique L is called the *Lebesgue integral of f on A*, and is denoted

$$\int_A f dm.$$

22. Simple Functions

We would like to have a criterion of Lebesgue integrability which avoids mention of the limit $L = \int_A f dm$, since the definition gives no clue as to how to find L. The relevant result for Riemann integrability (see Theorem 1.6) involves approximating f above and below by step functions. The significance of step functions is that their Riemann integrals (Proposition 1.5) are of the same form as Riemann sums. In our present situation we desire uncomplicated functions whose integrals are sums comparable to Lebesgue sums. It is not surprising that simple functions work (see section 19). Of course simple functions are bounded and measurable; they are also Lebesgue integrable.

22.1 Proposition: If $g:A \to \mathfrak{R}$ is simple, then g is Lebesgue integrable on A. Furthermore, if $g = \sum_{i=1}^{n} b_i \chi_{B_i}$ is any representation of g, then

$$\int_A g\,dm = \sum_{i=1}^{n} b_i \cdot m(B_i).$$

Proof: Note that we may assume without loss of generality that $b_i \neq b_j$ for $i \neq j$. (Otherwise, we could take unions of sets whose b_i's were the same. Since $m(B_i \cup B_j) = m(B_i) + m(B_j)$ for $i \neq j$, the result would be the same.) Now, given $\epsilon > 0$, let $0 < \delta \leq \epsilon/m(A)$ be so small that $\delta < |b_i - b_j|$ whenever $i \neq j$. If $P = (y_0, y_1, \ldots y_k)$ is a partition of $[\ell, u]$ (as in Definition 21.1) such that $\|P\| < \delta$, then at most one b_i can lie in each $[y_{j-1}, y_j)$. Therefore, for any

$$L(g,P) = \sum_{j=1}^{k} y_j^* \cdot m(\{x \in A \,|\, y_{j-1} \leqslant g(x) < y_j\}),$$

each set $\{x \in A \,|\, y_{j-1} \leqslant g(x) < y_j\}$ is either B_i (if $b_i \in [y_{j-1}, y_j)$), or $\emptyset$ (if no b_i is in $[y_{j-1}, y_j)$). Therefore, eliminating the zero terms in which $\emptyset$ occurs, there is a 1-1 correspondence between terms

$$b_i \cdot m(B_i)$$

and terms

$$y_j^* \cdot m(B_i),$$

where

$$b_i \in [y_{j-1}, y_j).$$

Thus

$$\left| L(g,P) - \sum_{i=1}^{n} b_i \cdot m(B_i) \right| = \left| \sum_{i=1}^{n} (y_j^* - b_i) \cdot m(B_i) \right|$$

$$\leqslant \sum_{i=1}^{n} |y_j^* - b_i| \cdot m(B_i).$$

But, if $b_i \in [y_{j-1}, y_j)$, then

$$|y_j^* - b_i| < |y_j - y_{j-1}| < \delta,$$

so that

$$\left| L(g,P) - \sum_{i=1}^{n} b_i \cdot m(B_i) \right| < \delta \left(\sum_{i=1}^{n} m(B_i) \right) = \delta \cdot m(A) \leqslant \epsilon.$$

Notice that the proposition implies that the sum $\sum_{i=1}^{n} b_i \cdot m(B_i)$ is the same no matter what representation is chosen for g. It is an important and useful fact that the Lebesgue integral, restricted to simple functions, satisfies certain nice properties expected of integrals. We will later extend these to other functions.

22.2 Proposition: If f and g are simple on A, then

(1) $\int_A f dm \leqslant \int_A g dm$ whenever $f \leqslant g$,
(2) $\int_A (f + g) dm = \int_A f dm + \int_A g dm$,
(3) $\int_A c f dm = c \int_A f dm$, for $c \in \mathbb{R}$.

Proof: Exercise 26.9.   □

22.3 Proposition: If f is simple on $A \cup B$, where A and B are disjoint bounded measurable sets, then

$$\int_{A \cup B} f dm = \int_A f dm + \int_B f dm.$$

Proof: Exercise 26.10.   □

23. Integrability of Bounded Measurable Functions

Now we state the criterion of Lebesgue integrability promised earlier.

23.1 Theorem: A bounded measurable function f is Lebesgue integrable on a bounded measurable set A if and only if for every $\epsilon > 0$, there exist simple functions f_1 and f_2 such that $f_1 \leqslant f \leqslant f_2$ and

$$\int_A f_2 dm - \int_A f_1 dm < \epsilon.$$

Proof: If f_1 and f_2 are simple and $f_1 \leqslant f \leqslant f_2$, then monotonicity (Proposition 22.2 (1)) guarantees that $\int_A f_1 dm \leqslant \int_A f_2 dm$. Since we should also expect that the integral of f, if it is to exist, should satisfy $\int_A f_1 dm \leqslant \int_A f dm \leqslant \int_A f_2 dm$, it would be natural to define

$$L = \text{lub} \left\{ \int_A f_1 dm \,\middle|\, f_1 \text{ simple and } f_1 \leqslant f \right\}.$$

This set is non-empty and bounded above (verify!) By hypothesis, if $\epsilon > 0$, then there are simple functions $f_1 \leqslant f \leqslant f_2$ such that

$$\int_A f_2 dm - \int_A f_1 dm < \epsilon/2.$$

For such simple functions, we have

$$\int_A f_1 dm \leqslant L \leqslant \int_A f_2 dm \qquad \text{(why?)}.$$

Now let $\delta = \epsilon/2m(A)$. All that remains to be proved is the following claim; the reader can verify that this will imply that f is Lebesgue integrable on A.

Claim: If $P = (y_0, y_1, \ldots, y_n)$ is a partition of $[\ell, u]$ (as defined in Definition 21.1) with $\|P\| < \delta$, then for any Lebesgue sum $L(f, P)$,

$$\int_A f_1 dm - \epsilon/2 \leqslant L(f, P) \leqslant \int_A f_2 dm + \epsilon/2.$$

Proof of Claim: On $A_i = \{x \in A \,|\, y_{i-1} \leqslant f(x) < y_i\}$,

$$f_2(x) \geqslant f(x) \geqslant y_{i-1} \geqslant y_i^* - \delta.$$

and

$$f_1(x) \leqslant f(x) < y_i \leqslant y_i^* + \delta.$$

Therefore,

$$f_1(x) - \delta \leqslant y_i^* \leqslant f_2(x) + \delta.$$

Hence, on integrating over A_i, by Proposition 22.2,

$$\int_{A_i} f_1 dm - \delta m(A_i) \leqslant y_i^* m(A_i) \leqslant \int_{A_i} f_2 dm + \delta m(A_i).$$

Adding these inequalities, for $i = 1, 2, \ldots, n$, we obtain by Proposition 22.3 and additivity of m:

$$\int_A f_1 dm - \delta m(A) \leqslant \sum_{i=1}^{n} y_i^* m(A_i) \leqslant \int_A f_2 dm + \delta m(A),$$

or

$$\int_A f_1 dm - \epsilon/2 \leqslant L(f, P) \leqslant \int_A f_2 dm - \epsilon/2.$$

To prove the converse, all we need to know is that f is measurable and bounded. In this case, if $\epsilon > 0$, let $P = (y_0, y_1, \ldots, y_n)$ be any partition of $[\ell, u]$ such that $\|P\| < \delta = \epsilon/m(A)$. Then if

$$A_i = \{x \in A \,|\, y_{i-1} \leqslant f(x) < y_i\},$$

and if

$$f_1 = \sum_{i=1}^{n} y_{i-1} \chi_{A_i} \quad \text{and} \quad f_2 = \sum_{i=1}^{n} y_i \chi_{A_i},$$

then we have $f_1 \leqslant f \leqslant f_2$, and

$$\int_A f_2 \, dm - \int_A f_1 \, dm = \sum_{i=1}^{n} (y_i - y_{i-1}) m(A_i) < \delta \left(\sum_{i=1}^{n} m(A_i) \right) = \epsilon. \quad \square$$

The proof of the theorem yields several important corollaries. The first is a little surprising; it says that by restricting ourselves to *measurable* bounded functions in the definition of Lebesgue integrability, we have already eliminated all non-Lebesgue integrable functions.

23.2 Corollary: If f is bounded and measurable on a bounded measurable set A, then f is Lebesgue integrable on A.

Proof: See the proof of the theorem. $\quad \square$

Note that already we can see that many non-Riemann integrable functions are Lebesgue integrable. (Name some.)

The next corollary is not so astounding, but it is of theoretical importance. In other treatments of the subject, it is often used as a definition for the Lebesgue integral. We will make use of it ourselves to extend our definition to unbounded measurable functions (see section 25.)

23.3 Corollary: If f is bounded and measurable on a bounded measurable set A, then

$$\int_A f \, dm = \text{lub} \left\{ \int_A f_1 \, dm \,\middle|\, f_1 \text{ is simple and } f_1 \leqslant f \right\}$$

$$= \text{glb} \left\{ \int_A f_2 \, dm \,\middle|\, f_2 \text{ is simple and } f_2 \geqslant f \right\}.$$

Proof: See Exercise 26.12. $\quad \square$

23.4 Corollary: If $m(A) = 0$ and f is bounded and measurable on A, then

$$\int_A f \, dm = 0.$$

Proof: See Exercise 26.13. □

24. Elementary Properties of the Integral for Bounded Functions

The Lebesgue integral shares with the Riemann integral the nice properties stated in the following theorem.

24.1 Theorem: If f and g are bounded and measurable on the bounded measurable set A, then:

(1) (Monotonicity) If $f \leqslant g$, then $\int_A f dm \leqslant \int_A g dm$;

(2) If there are real numbers ℓ and u such that $\ell \leqslant f \leqslant u$, then
$$\ell \cdot m(A) \leqslant \int_A f dm \leqslant u \cdot m(A);$$

(3) (Linearity) $\int_A cf dm = c \int_A f dm$ for c real;

(4) (Linearity) $\int_A (f + g) dm = \int_A f dm + \int_A g dm$;

(5) $|\int_A f dm| \leqslant \int_A |f| dm$.

Proof: (1) If $f_1 \leqslant f$, then $f_1 \leqslant g$. Hence by Corollary 23.3,

$$\int_A f dm = \text{lub} \left\{ \int_A f_1 dm \, \middle| \, f_1 \leqslant f, f_1 \text{ simple} \right\}$$
$$\leqslant \text{lub} \left\{ \int_A g_1 dm \, \middle| \, g_1 \leqslant g, g_1 \text{ simple} \right\} = \int_A g dm.$$

(2) Use (1), noting that $g(x) = \ell$ and $h(x) = u$ are simple functions, so that their integrals over A are easy to compute.

(3) Follows from the identities (which the reader should verify)

$$c \cdot \text{lub} \left\{ \int_A f_1 dm \, \middle| \, f_1 \leqslant f, f_1 \text{ simple} \right\} = \text{lub} \left\{ \int_A cf_1 dm \, \middle| \, f_1 \leqslant f, f_1 \text{ simple} \right\}$$
$$= \text{lub} \left\{ \int_A g_1 dm \, \middle| \, g_1 \leqslant cf, g_1 \text{ simple} \right\}.$$

Note that cf is measurable and bounded on A if f is.

(4) Recall that $f + g$ is measurable and bounded on A, hence Lebesgue integrable. Let $\epsilon > 0$. By virtue of Corollary 23.3, there are simple functions $f_1 \leqslant f$, $g_1 \leqslant g$, such that

$$\int_A f_1 dm > \int_A f dm - \epsilon \qquad \text{and} \qquad \int_A g_1 dm > \int_A g dm - \epsilon.$$

Therefore,

$$\int_A (f + g) dm \underset{(1)}{\geqslant} \int_A (f_1 + g_1) dm \underset{22.2}{=} \int_A f_1 dm + \int_A g_1 dm$$
$$> \int_A f dm + \int_A g dm - 2\epsilon.$$

Since ϵ was arbitrary, we have

$$\int_A (f+g)dm \geqslant \int_A fdm + \int_A gdm.$$

The reverse inequality is proved similarly, using the fact that

$$\int_A fdm = \text{glb}\left\{\int_A f_2 dm \,\middle|\, f_2 \geqslant f, \; f_2 \text{ simple}\right\}.$$

(Exercise 26.14.)

(5) Note that $|f|$ is bounded and measurable. Also $f \leqslant |f|$, and $-f \leqslant |f|$. Hence (5) follows from (1) and (3).   □

Another fundamental property which the Riemann integral has is additivity on intervals:

$$\int_a^b f(x)dx + \int_b^c f(x)dx = \int_a^c f(x)dx.$$

Of course, for the Lebesgue integral we are no longer restricted to closed intervals, but may integrate over arbitrary bounded measurable sets. Our version of additivity might be

$$\int_A fdm + \int_B fdm = \int_C fdm,$$

where $C = A \cup B$ disjointly (actually, the intersection could contain a set of measure 0, as we will see). It turns out that an even stronger kind of additivity, countable additivity, holds. But first the simpler property.

24.2 Theorem: If A and B are disjoint bounded measurable sets, and $f:A \cup B \to \mathcal{R}$ is bounded and measurable, then

$$\int_{A \cup B} fdm = \int_A fdm + \int_A fdm.$$

Proof: Note that f is bounded and measurable on each of A, B, so that the integrals all exist. Now let g_1 be simple on A such that $g_1(x) \leqslant f(x)$ for all $x \in A$, and let h_1 be simple on B such that $h_1(x) \leqslant f(x)$ for all $x \in B$. Then the function f_1 defined by

$$f_1(x) = \begin{cases} g_1(x) & \text{if } x \in A \\ h_1(x) & \text{if } x \in B, \end{cases}$$

is simple on $A \cup B$, and $f_1 \leqslant f$ on $A \cup B$. Furthermore, by additivity for simple functions (Proposition 22.3),

$$\int_A g_1 dm + \int_B h_1 = \int_A f_1 dm + \int_B f_1 dm$$

$$\underset{22.3}{=} \int_{A \cup B} f_1 dm \underset{24.1}{\leqslant} \int_{A \cup B} f dm.$$

Therefore, taking lub's, we obtain

$$\int_A f dm + \int_B f dm \leqslant \int_{A \cup B} f dm$$

(Corollary 23.3). On the other hand, if f_1 is simple and if $f_1 \leqslant f$ on $A \cup B$, then $f_1 \leqslant f$ on A and on B separately, so that

$$\int_{A \cup B} f_1 dm = \int_A f_1 dm + \int_B f_1 dm \leqslant \int_A f dm + \int_B f dm.$$

Again taking lub's, we get $\int_{A \cup B} f dm \leqslant \int_A f dm + \int_B f dm$. $\qquad \square$

24.3 Theorem: (Countable additivity for the integral) Let A be a bounded measurable set. If $f:A \to \Re$ is bounded and measurable, and if $A = \overset{\infty}{\underset{i=1}{\cup}} A_i$, where the A_i are pairwise disjoint measurable sets, then

$$\int_A f dm = \overset{\infty}{\underset{i=1}{\Sigma}} \int_{A_i} f dm.$$

Proof: We need to show that $|\int_A f dm - \overset{n}{\underset{i=1}{\Sigma}} \int_{A_i} f dm|$ becomes arbitrarily small as $n \to \infty$. But if $B_n = \overset{\infty}{\underset{i=n+1}{\cup}} A_i$, finite additivity (Theorem 24.2) and induction give

$$\left| \int_A f dm - \overset{n}{\underset{i=1}{\Sigma}} \int_{A_i} f dm \right| \underset{24.2}{=} \left| \int_A f dm - \int_{\underset{i=1}{\overset{n}{\cup}A_i}} f dm \right|$$

$$\underset{24.2}{=} \left| \int_{B_n} f dm \right| \underset{24.1}{\leqslant} \int_{B_n} |f| dm.$$

Given $\epsilon > 0$, there is a simple function f_1 on A with $0 \leqslant f_1 \leqslant |f|$ on A, such that $\int_A f_1 dm > \int_A |f| dm - \epsilon/2$ (Corollary 23.3). It follows that for all n,

$$\int_{B_n} f_1 dm > \int_{B_n} |f| dm - \epsilon/2.$$

(Exercise 26.16). But f_1 is bounded; say $f_1(x) \leqslant u$ for $x \in A$. Hence by Theorem 24.1,

$$\int_{B_n} f_1 dm \leqslant u \cdot m(B_n).$$

Since $\{B_n\}$ is a decreasing sequence of measurable sets with empty inter-section, $\lim_{n \to \infty} m(B_n) = 0$ (Corollary 10.10). Thus for large enough n, $u \cdot m(B_n) < \epsilon/2$. Therefore,

$$\left| \int_A f dm - \sum_{i=1}^{n} \int_{A_i} f dm \right| \leq \int_{B_n} |f| dm < \int_{B_n} f_1 dm + \epsilon/2$$

$$\leq u \cdot m(B_n) + \epsilon/2 < \epsilon. \qquad \square$$

If we define $\mu(B) = \int_B f dm$ for some bounded measurable function $f:A \to \mathcal{R}$, where $B \subset A$, and A is bounded and measurable, then μ is a set function defined on all measurable subsets of A. It is countably additive by Theorem 24.3. If $f \geq 0$, the μ is a measure (Definition 6.1 and Exercise 26.17). If $f(x) = 1$ for all $x \in A$, then $\mu = m$, the Lebesgue measure.

As a direct consequence of additivity, we have a couple of important and easy results which say that sets of measure 0 don't matter in integra-tion.

24.4 Proposition: If A is a bounded measurable set, $B \subset A$ has $m(B) = 0$, and f is bounded and measurable on A, then $\int_A f dm = \int_{A \setminus B} f dm$.

Proof: Exercise 26.18. $\qquad \square$

24.5 Proposition: If f is bounded measurable on A, and g is measurable on A, and $f = g$ a.e. on A, then $\int_A f dm = \int_A g dm$.

Proof: Strictly speaking, g may be unbounded, so that $\int_A g dm$ is undefined. In section 25, we will extend the definition of the integral to cover unbounded measurable functions. In the meantime, the previous proposition suggests how to deal with a function like g, which is "almost bounded." See Exercise 26.19 for details. $\qquad \square$

The last two propositions can be used to extend previous results to slightly more general situations. For example, in Theorem 24.2, if we assume $m(A \cap B) = 0$, rather than the stronger property of disjointness, then still $\int_{A \cup B} f dm = \int_A f dm + \int_B f dm$.

25. The Lebesgue Integral for Unbounded Functions

The definition of the Lebesgue integral given in section 21 (Defini-tion 21.1) will not work for unbounded functions, since $\{f(x) | x \in A\}$

will not have both a finite lub and a finite glb. To extend the definition to unbounded functions, therefore, we must look further. We choose to take Corollary 23.3 as our starting point. It says that for a bounded measurable function f on a bounded measurable set A,

$$\int_A f dm = \text{lub}\left\{\int_A f_1 dm \,\middle|\, f_1 \text{ simple and } f_1 \leqslant f\right\}$$

$$= \text{glb}\left\{\int_A f_2 dm \,\middle|\, f_2 \text{ simple and } f_2 \geqslant f\right\}.$$

We are going to take this *property* which holds for bounded measurable functions, and use it as the basis for a *definition* for unbounded measurable functions. However, there are difficulties. Namely, if f is unbounded above, there will be no simple functions f_2 with $f_2 \geqslant f$; and if f is unbounded below, there will be no simple functions f_1 with $f_1 \leqslant f$. That is, the sets in question could be empty, and the lub and glb may not make sense (may not exist as real numbers). The solution is to consider first only non-negative functions f. For such functions, there are always simple f_1 with $f_1 \leqslant f$.

25.1 Definition: Let A be bounded and measurable, $f{:}A \rightarrow \mathcal{R}$ measurable and *non-negative*. Define $\int_A f dm = \text{lub}\{\int_A f_1 dm \,|\, f_1 \text{ simple and } f_1 \leqslant f\}$.

Note that if $\{\int_A f_1 dm \,|\, f_1 \text{ simple and } f_1 \leqslant f\}$ is unbounded above, then the lub is not a real number. In that case, we take the lub to be ∞; that is, we say $\int_A f dm = \infty$.

Because of Corollary 23.3, this definition is consistent with the definition of the Lebesgue integral for *bounded* measurable functions. Perhaps it has occurred to you that there is no necessity to restrict the definition to *measurable* non-negative functions. The set $\{\int_A f_1 dm \,|\, f_1$ simple and $f_1 \leqslant f\}$ is non-empty in that case as well. The reason for the restriction is that measurability is required to prove some of the nice limit properties of the Lebesgue integral which appear in the next chapter.

Now, to deal with an arbitrary measurable function f, we split it into positive and negative parts: let

$$f_+(x) = \begin{cases} f(x) & \text{if } f(x) \geqslant 0 \\ 0 & \text{if } f(x) < 0. \end{cases}$$

$$f_-(x) = \begin{cases} -f(x) & \text{if } f(x) \leqslant 0 \\ 0 & \text{if } f(x) > 0. \end{cases}$$

Then it is immediate that

$$f_+ = \max\{f,0\} = \frac{1}{2}(|f| + f) \geqslant 0,$$

$$f_- = \max\{-f,0\} = \frac{1}{2}(|f| - f) \geqslant 0,$$

$$f = f_+ - f_-, \quad \text{and} \quad |f| = f_+ + f_-.$$

Therefore, we might expect $\int_A f dm = \int_A f_+ dm - \int_A f_- dm$. The only difficulty arises if $\int_A f_+ dm = \int_A f_- dm = \infty$, since $\infty - \infty$ cannot reasonably be given a value.

25.2 Definition: Let A be a bounded measurable set, and $f:A \to \mathcal{R}$ measurable. Then f is *integrable* on A if at least one of $\int_A f_+ dm$ or $\int_A f_- dm$ is finite, and f is *summable* on A if both $\int_A f_+ dm$ and $\int_A f_- dm$ are finite. In either case, define $\int_A f dm = \int_A f_+ dm - \int_A f_- dm$.

Of course, in the definition, $\infty - c$ is called ∞, and $c - \infty$ is called $-\infty$ for every (finite) real number c. Linearity of the Lebesgue integral for bounded functions (Theorem 24.1) shows that this definition is still consistent with the one for bounded functions.

As one would hope, the nice properties proved in section 24 for bounded functions carry over to this new situation. The proofs remain the same, except as indicated. All references to Corollary 23.3 now are to Definition 25.1 and 25.2, of course. Let us introduce the notation

$$\mathcal{L}(A) = \{f:A \to \mathcal{R} \mid f \text{ is summable on } A\},$$

for A bounded and measurable.

25.3 Theorem: (Additivity) If A and B are disjoint, bounded measurable sets and $f \in \mathcal{L}(A \cup B)$, then $f \in \mathcal{L}(A) \cup \mathcal{L}(B)$, and $\int_{A \cup B} f dm = \int_A f dm + \int_B f dm$.

Proof: By the proof of Theorem 24.2,

$$\int_{A \cup B} f_+ dm = \int_A f_+ dm + \int_B f_+ dm$$

and

$$\int_{A \cup B} f_- dm = \int_B f_- dm + \int_B f_- dm.$$

The result follows by subtraction (Definition 25.2).   □

25.4 Theorem: Let A be bounded and measurable, $f, g \in \mathcal{L}(A)$. Then

(1) (Monotonicity) If $f \leqslant g$, then $\int_A f dm \leqslant \int_A g dm$;
(2) If there are real numbers ℓ and u such that $\ell \leqslant f \leqslant u$, then

$$\ell \cdot m(A) \leqslant \int_A f dm \leqslant u \cdot m(A);$$

(3) (Linearity) $cf \in \mathcal{L}(A)$ for every real number c, and

$$\int_A cf dm = c \int_A f dm:$$

(4) (Linearity) $f + g \in \mathcal{L}(A)$, and $\int_A (f + g) dm = \int_A f dm + \int_A g dm$;
(5) $|f| \in \mathcal{L}(A)$ and $|\int_A f dm| \leqslant \int_A |f| dm$.

Proof: (1) Clearly $f_+ \leqslant g_+$ and $f_- \geqslant g_-$. Apply the proof of Theorem 24.1 (1), and Definition 25.2.

(2) By (1) of this theorem. Note that since f is bounded in this case, part (2) of this theorem is the same as part (2) of Theorem 24.1.

(3) Apply the proof of Theorem 24.1 to f_+ and f_-.

(5) Let $B = \{x \in A \mid f(x) \geqslant 0\}$, and $C = \{x \in A \mid f(x) < 0\}$. Then by additivity (Theorem 25.3),

$$\int_A |f| dm = \int_B |f| dm + \int_C |f| dm$$
$$= \int_B f_+ dm + \int_C f_- dm = \int_A f_+ dm + \int_A f_- dm,$$

and both of the latter integrals are finite by assumption (see Definition 25.2). The remainder of the proof of this part is as in part (5) of Theorem 24.1.

(4) The proof of the corresponding part of Theorem 24.1 does not apply here, even for $f \geqslant 0$ and $g \geqslant 0$, since we used

$$\int_A f dm = \text{glb} \left\{ \int_A f_2 dm \mid f_2 \text{ simple and } f_2 \geqslant f \right\},$$

which is not valid for unbounded f. The simplest proof in our present case involves using the Monotone Convergence Theorem of section 28. No circularity will result from using that theorem here.

Let us write $h = f + g$, and

$$F = \{x \in A \mid f(x) \geqslant 0\}, \quad F' = A \backslash F;$$

$$G = \{x \in A \mid g(x) \geqslant 0\}; \quad G' = A \backslash G;$$

$$H = \{x \in A \mid h(x) \geqslant 0\}; \quad H' = A \backslash H.$$

Note that $A = (F \cap G) \cup (F \cap G') \cup (F' \cap G) \cup (F' \cap G')$ disjointly. We show the result holds on each of these intersections, then apply additivity (Theorem 25.3).

(a) On $F \cap G$, by Exercise 20.32 there are non-decreasing sequences of simple functions $\{f_n\}$, $\{g_n\}$ converging pointwise to f, g respectively. Then the sequence $\{f_n + g_n\}$ is non-decreasing and converges pointwise to $h = f + g$. This Monotone Convergence Theorem says that

$$\lim_{n \to \infty} \int_{F \cap G} f_n dm = \int_{F \cap G} f dm,$$

$$\lim_{n \to \infty} \int_{F \cap G} g_n dm = \int_{F \cap G} g dm,$$

and

$$\lim_{n \to \infty} \int_{F \cap G} (f_n + g_n) dm = \int_{F \cap G} h dm.$$

But since f_n and g_n are simple functions, Theorem 24.1 or Theorem 22.2 gives

$$\int_{F \cap G} (f_n + g_n) dm = \int_{F \cap G} f_n dm + \int_{F \cap G} g_n dm.$$

It follows easily that $\int_{F \cap G} h dm = \int_{F \cap G} f dm + \int_{F \cap G} g dm$.

(b) The result $\int_{F' \cap G'} h dm = \int_{F' \cap G'} f dm + \int_{F' \cap G'} g dm$ follows by considering $-f$, $-g$, and $-h$ on $F' \cap G'$, and applying the technique of (a).

(c) On $F \cap G'$, consider the disjoint representation

$$F \cap G' = (F \cap G' \cap H) \cup (F \cap G' \cap H')$$

On $F \cap G' \cap H$, we have $f \geqslant 0$, $-g > 0$, and $h \geqslant 0$, so that $f = h + (-g)$. Using the technique of (a), and part (3) of this theorem, we obtain

$$\int_{F \cap G' \cap H} f dm = \int_{F \cap G' \cap H} h dm - \int_{F \cap G' \cap H} g dm.$$

Similarly, on $F \cap G' \cap H'$, $-g = -h + f$, and $-g$, $-h$, f are all non-negative. Therefore our previous results give

$$-\int_{F \cap G' \cap H'} g\, dm = -\int_{F \cap G' \cap H'} h\, dm + \int_{F \cap G' \cap H'} f\, dm.$$

Combining, we get $\int_{F \cap G'} f\, dm + \int_{F \cap G'} g\, dm = \int_{F \cap G'} h\, dm$.

(d) Similarly, $\int_{F' \cap G} f\, dm + \int_{F' \cap G} g\, dm = \int_{F' \cap G} h\, dm$. The sum of (a) - (d) gives the result; all the numbers (integrals) concerned are finite, so $f + g \in \mathcal{L}(A)$. ☐

25.5 Theorem: (Countable Additivity) Let A be bounded and measurable, $f \in \mathcal{L}(A)$, $A = \bigcup_{i=1}^{\infty} A_i$, where the A_i are pairwise disjoint measurable sets. Then $\int_A f\, dm = \sum_{i=1}^{\infty} \int_{A_i} f\, dm$.

Proof: By the proof of Theorem 24.3, $\int_A f_+ dm = \sum_{i=1}^{\infty} \int_{A_i} f_+ dm$ and $\int_A f_- dm = \sum_{i=1}^{\infty} \int_{A_i} f_- dm$. Therefore,

$$\int_A f\, dm = \int_A f_+ dm - \int_A f_- dm$$

$$= \sum_{i=1}^{\infty} \int_{A_i} f_+ dm - \sum_{i=1}^{\infty} \int_{A_i} f_- dm$$

$$= \sum_{i=1}^{\infty} \left(\int_{A_i} f_+ dm - \int_{A_i} f_- dm \right) = \sum_{i=1}^{\infty} \int_{A_i} f\, dm.$$

☐

25.6 Corollaries: (1) If A is bounded and measurable, and $B \subset A$ has measure 0, and $f \in \mathcal{L}(A)$, then $\int_A f\, dm = \int_{A \setminus B} f\, dm$.

(2) If $f \in \mathcal{L}(A)$ and $g = f$ a.e. on A, then $\int_A f\, dm = \int_A g\, dm$. ☐

The definition of the Lebesgue integral for unbounded functions may be approached in many ways other than the one we chose. You should be able, using the results of the next chapter especially, to prove that other definitions you may encounter are equivalent to ours.

26. Exercises

26.1 Prove uniqueness of the Lebesgue integral.

26.2 Find two Lebesgue sums of $f(x) = x^2$ on $[-1,2]$, relative to the partition $(0, 1/4, 1/2, 2/3, 1, 2, 4, 5)$ of $[\ell,u]$.

26.3 Show that if f is monotone on $[a,b]$ every Riemann sum of f is a Lebesgue sum. Is the converse true?

26.4 Where does boundedness of f come into Definition 21.1? Where does measurability of f come in?

26.5 Show directly from Definition 21.1 that if f is bounded and $m(A) = 0$, then $\int_A f\,dm = 0$.

26.6 As an easy case of Proposition 22.1, show from the definition that if $f(x) = c$ for $x \in A$, then $\int_A f\,dm = c \cdot m(A)$.

26.7 Show directly from Definition 21.1 that if $f(x) = x$ for $x \in [0,1]$, then $\int_{[0,1]} f\,dm = \frac{1}{2}$.

26.8 Find $\int_{[0,1]} \chi_Q\,dm$.

26.9 Prove Proposition 22.2. (Hint: if $f = \sum\limits_{i=1}^{k} c_i \chi_{C_i}$ and $g = \sum\limits_{j=1}^{n} d_j \chi_{D_j}$, then write $f = \sum\limits_{i=1}^{k} \sum\limits_{j=1}^{n} c_i \chi_{C_i \cap D_j}$, etc.)

26.10 Prove Proposition 22.3.

26.11 Finish the proof of Theorem 23.1; show that the *Claim* implies that f is Lebesgue integrable on A.

26.12 Prove Corollary 23.3.

26.13 Prove Corollary 23.4.

26.14 Prove the remainder of Theorem 24.1(4); $\int_A f\,dm + \int_A g\,dm \geqslant \int_A (f + g)\,dm$.

26.15 Let $A = \bigcup\limits_{i=1}^{\infty} A_i$, where the A_i are disjoint and measurable and A is bounded. If f is measurable on each A_i, show that f is measurable on A. Does the fact that f is bounded on each A_i imply that f is bounded on A?

26.16 (a) If $0 \leqslant g$ is bounded and measurable on A, and $B \subset A$ is measurable, show that $\int_A g\,dm \geqslant \int_B g\,dm$.
(b) If $f \leqslant g$ are bounded and measurable, $B \subset A$ is measurable, and $\int_A f\,dm > \int_A g\,dm - \epsilon$, then show that $\int_B f\,dm > \int_B g\,dm - \epsilon$ without using the linearity of the integral.

26.17 If $\mu(B) = \int_B f\,dm$, for $f \geqslant 0$ and bounded and measurable on a bounded and measurable set A, show that μ is a measure on the measurable subsets of A.

26.18 Prove Proposition 24.4.

26.19 Prove Proposition 24.5.

26.20 (a) Prove that if f is bounded and measurable on $A \cup B$, where A and B are bounded and measurable, then
$$\int_{A \cup B} f \, dm + \int_{A \cap B} f \, dm = \int_A f \, dm + \int_B f \, dm.$$
(b) If $m(A \cap B) = 0$, show that $\int_{A \cup B} f \, dm = \int_A f \, dm + \int_B f \, dm.$

26.21 Prove that for any measurable function f, if $m(A) = 0$, then $\int_A f \, dm = 0$.

26.22 Complete the proof of Theorem 25.4 (1), (2), (3), and (5).

26.23 (a) Suppose that $f \geqslant 0$ and $\int_A f \, dm = 0$. Show that $f = 0$ a.e. on A. (Hint: Let $B = \{x \in A \, | \, f(x) > 0\} = \bigcup_{n=1}^{\infty} \{x \in A \, | \, f(x) > 1/n\}$. If $m(B) > 0$, obtain a contradiction.)
(b) Let $f \geqslant g$ and $\int_A f \, dm = \int_A g \, dm$. Show that $f = g$ a.e. on A.

26.24 If f is measurable function such that $\int_B f \, dm = 0$ for all measurable sets $B \subset A$, show that $f = 0$ a.e. on A.

26.25 If f is measurable on $[a,b]$ and $\int_{[a,c]} f \, dm = 0$ for all $c < b$, show that $f = 0$ a.e. on $[a,b]$.

26.26 Prove that if f is measurable on A and $|f| \leqslant g$ and $g \in \mathcal{L}(A)$, then $f \in \mathcal{L}(A)$.

26.27 If $f \in \mathcal{L}(A)$ and B is a measurable subset of A, show that $f \in \mathcal{L}(B)$.

26.28 If $f \in \mathcal{L}(A)$ and $g(x) = f(x) \sin(ax)$, show that $g \in \mathcal{L}(A)$.

26.29 Determine whether $f(x) = 1/x$ is Lebesgue integrable on $(0,1)$. Is it summable? Find $\int_{(0,1)} f \, dm$.

26.30 Find $\int_{(0,1)} g \, dm$ if $g(x) = x^{-\frac{1}{2}}$.

26.31 Show that if $f \in \mathcal{L}(A)$ and $B \in A$ is measurable, then $\chi_B f \in \mathcal{L}(A)$ and
$$\int_A (\chi_B f) \, dm = \int_B f \, dm.$$

26.32 If $g \in \mathcal{L}(A)$ and $A \supset E_1 \supset E_2 \supset \cdots$, and $\bigcap_{n=1}^{\infty} E_n = \emptyset$, show that
$$\lim_{n \to \infty} \int_{E_n} g \, dm = 0.$$

26.33 If $f \in \mathcal{L}(A)$, and $\epsilon > 0$, show that there exists a $\delta > 0$ such that whenever $D \subset A$ and $m(D) < \delta$, then $|\int_D f \, dm| < \epsilon$. (Hint: Let

$$A_n = \{x \in A \mid |f(x)| < n\}.$$

Then find N such that $|\int_{A \backslash A_n} f \, dm| < \epsilon/2$. Let $\delta = \epsilon/2N$.)

26.34 Let g, $h \in \mathcal{L}(A)$ and let $a \leqslant g(x) \leqslant b$ for a.e. $x \in A$. Show that there is a number $c \in [a,b]$ such that $\int_A g |h| \, dm = c(\int_A |h| \, dm)$.

26.35 In Riemann integration, the change of variable formula provides the following equation for g continuous:

$$\int_a^b g(u) \, du = \int_{c-b}^{c-a} g(c-t) \, dt. \qquad (\text{Let } t = c - u.)$$

We can put this in Lebesgue terms as follows: given g summable on A, let $B = c - A = \{c - x \mid x \in A\}$, where c is a fixed real number Let $h(x) = g(c-x)$, $x \in B$. Prove that h is summable on B, and

$$\int_A g \, dm = \int_B h \, dm.$$

(Hint: consider $h_+(x) = g_+(c-x)$ first, and let $h_1 \leqslant h_+$ be simple. Let $g_1(x) = h_1(c-x)$. Then g_1 is simple and $g_1 \leqslant g_+$. Show that $\int_A g_1 \, dm = \int_B h_1 \, dm$ (see Lemma 7.6 and Exercise 9.23). Conclude that $\int_A g \, dm \geqslant \int_B h \, dm$. Then prove the opposite inequality.)

26.36 Let $g \in \mathcal{L}[a,b]$, $h(x) = \int_{[a,x]} g \, dm$, for $x \in [a,b]$. Show that h is continuous on $[a,b]$. (Hint: consider first the case where g is simple.)

26.37 Prove the following generalization of Exercise 26.39. If $g \in \mathcal{L}(A)$ and $g \geqslant 0$, then for any $\epsilon > 0$, there is a $\delta > 0$ such that whenever $D \subset A$ is measurable and $m(D) < \delta$, we have $\int_D g \, dm < \epsilon$. (Hint: Consider $A_n = \{x \in A \mid g(x) \leqslant n\}$. There exists an n such that $\int_{A \backslash A_n} g \, dm < \epsilon/2$.)

CHAPTER 6

Convergence and The Lebesgue Integral

27. Examples

One of our reasons for being dissatisfied with the Riemann integral had to do with convergence properties. We are interested in results concerning the integral of the limit of a sequence of functions. In particular, under what conditions will $\lim_{n \to \infty} \int_a^b f_n(x)dx = \int_a^b (\lim_{n \to \infty} f_n(x))dx$? It is known that if $f_n \to f$ uniformly on $[a,b]$ and all the f_n and f are Riemann integrable, then $\lim_{n \to \infty} \int_a^b f_n(x)dx = \int_a^b f(x)dx$. (See Exercises 5.25 and 31.1.)

There are numerous examples in which $f_n \to f$ pointwise on $[a,b]$, but $\lim_{n \to \infty} \int_a^b f_n(x)dx \neq \int_a^b f(x)dx$.

27.1 Example: Let

$$f_n(x) = \begin{cases} 2^n & \text{when } 1/2^n \leq x \leq 1/2^{n-1} \\ 0 & \text{otherwise} \end{cases}$$

for $n = 1,2,3, \cdots$ on $[0,1]$. Clearly $f_n \to 0$ pointwise, but

$$\lim_{n \to \infty} \int_0^1 f_n(x)dx = 1 \neq 0 = \int_0^1 (\lim_{n \to \infty} f_n(x))dx.$$

It is tempting to suppose that the requirement of uniform boundedness (i.e. $|f_n(x)| < M$ for all n,x) might remedy the situation, but the next example shows this is insufficient.

77

27.2 Example: For each $n = 1,2,3, \cdots$ let

$$f_{n,k} = \chi_{[\frac{k}{2^n}, \frac{k+1}{2^n}]}$$

for $k = 0,1, \cdots, 2^n - 1$ on $[0,1]$. Arrange the $f_{n,k}$ in a sequence $\{f_p\}$ first by order of n and then by k: $f_1 = f_{1,0}; f_2 = f_{1,1}; f_3 = f_{2,0}; f_4 = f_{2,1}; f_5 = f_{2,2}; f_6 = f_{2,3}; f_7 = f_{3,0}; \cdots$.

Then $\lim\limits_{p \to \infty} \int_0^1 f_p(x)dx = 0$ but the sequence $\{f_p\}$ does not converge (show this).

Not even monotonicity and the existence of a limit help. Recall Example 4.1 of Chapter 1 in which a (strictly increasing) uniformly bounded sequence of Riemann integrable functions converged to a function which was not even Riemann integrable.

We would hope that convergence properties for the Lebesgue integral might be better. However, this is not immediately clear. In Example 27.1 above, the f_n are all Lebesgue integrable (summable) and converge (pointwise) to a Lebesgue integrable function f. Similarly, in Example 27.2 all the $f_{n,k}$ are Lebesgue integrable (summable) and are uniformly bounded. However, they do not even converge almost everywhere. Thus at first glance the Lebesgue integral does not seem to offer much of an advantage.

However, it is worth noting that the difficulties presented by Example 4.1 of Chapter 1 are overcome by the Lebesgue theory. In fact, $\lim\limits_{n \to \infty} \int_{[0,1]} f_n(x)dm = 0 = \int_{[0,1]} f(x)dm$, now that the limit function χ_Q is Lebesgue integrable (summable). The advantage of the Lebesgue theory is that under much less stringent hypotheses than for the Riemann case, convergent sequences of integrable functions have integrable limits.

28. Convergence Theorems

There are two major convergence theorems involving the Lebesgue integral. They are the *Monotone Convergence Theorem* and the *Lebesgue Dominated Convergence Theorem*, and neither is true if we restrict our attention to Riemann integrable functions. Thus we will observe that our new Lebesgue theory actually offers an improvement over the Riemann theory with regard to convergence properties. It is this improvement which makes the Lebesgue theory valuable in many theoretical applications. We will use it in later chapters on Fourier analysis.

To prove the Monotone Convergence Theorem we need the following Lemma.

28.1 Lemma: Let g be a non-negative measurable function on a bounded, measurable set A. If $\{A_i\}_{i=1}^{\infty}$ are measurable subsets of A with

$$A_1 \subset A_2 \subset \cdots,$$

and if α is a real number satisfying $\alpha \geq \int_{A_n} g\,dm$ for all $n = 1,2, \cdots$. then $\alpha \geq \int_{\cup A_n} g\,dm$.

Proof: Exercise 31.3. $\qquad\qquad\qquad\qquad\qquad\qquad\qquad$ $\square$

28.2 Theorem (Monotone Convergence Theorem): Let A be a bounded measurable subset of $\mathcal{R}$ and $\{f_n\}$ be a sequence of measurable functions on A such that $0 \leq f_1 \leq f_2 \leq \cdots$. Let f be the (pointwise) limit of $\{f_n\}$. Then f is integrable and $\lim_{n \to \infty} \int_A f_n\,dm = \int_A f\,dm$.(*)

Proof: First, f is integrable since f is measurable (18.6) and $f \geq 0$. Since $f \leq g$ implies $\int_A f \leq \int_A g$ by monotonicity of the integral (25.4(1)), it follows that $\{\int_A f_n\,dm\}$ is an increasing sequence (possibly including ∞) and therefore has a limit L (possibly equal to ∞). Clearly $L \leq \int_A f\,dm$ since $f \geq f_n$ implies $\int_A f\,dm \geq \int_A f_n\,dm$ for all n.

To show $L \geq \int_A f\,dm$, let c be a number such that $0 < c < 1$ and let g be a simple function such that $0 \leq g \leq f$.

Define $A_n = \{x \mid f_n(x) \geq cg(x)\}$ for $n = 1,2, \cdots$. Clearly the A_n satisfy the hypotheses of the previous Lemma and

$$L = \lim_{n \to \infty} \int_A f_n\,dm \geq \int_{A_n} f_n\,dm \geq c \int_{A_n} g\,dm \text{ for any } n = 1,2 \cdots.$$

By the Lemma, $L \geq c\int_A g\,dm$ and since this is true for all $c \in (0,1)$, $L \geq \int_A g\,dm$. Therefore by the definition of the integral (25.1), $L \geq \int_A f\,dm$.
$\qquad\qquad\qquad\qquad\qquad\qquad\qquad\qquad\qquad\qquad\qquad\qquad\qquad$ $\square$

Since sets of measure zero have no effect on Lebesgue integrals, the following Corollary is immediate.

*This theorem is true even under an extended concept of "function" which allows infinite values. Under this interpretation, every monotone increasing sequence of functions will converge to some function.

If $f \geq 0$ and $f(x) = \infty$ for $x \in B$, then $\int_B f\,dm = \infty$ if $m(B) > 0$ and $\int_B f\,dm = 0$ if $m(B) = 0$.

28.3 Corollary: If A and $\{f_n\}$ are as in the Theorem and $f(x) = \lim\limits_{n \to \infty} f_n(x)$ *almost everywhere, then* $\lim\limits_{n \to \infty} \int_A f_n dm = \int_A f dm$. □

Another Corollary follows immediately from the close relationship between sequences and series.

28.4 Corollary: If $\{f_n\}_{n=1}^{\infty}$ is a sequence of non-negative measurable functions on a bounded measurable set A, then $\int_A \sum\limits_{n=1}^{\infty} f_n dm = \sum\limits_{n=1}^{\infty} \int_A f_n dm$.

Proof: The partial sums of the series form a monotone increasing sequence. □

The hypothesis that the f_n are non-negative seems a bit restrictive in the Monotone Convergence Theorem. What is really at stake is the desire to avoid the following kind of situation.

28.5 Example: Let

$$f_n(x) = \begin{cases} 0 \text{ on } (1/n,1] \\ -1/x \text{ on } (0,1/n] \\ 0 \text{ at } x = 0. \end{cases}$$

Each f_n is integrable (not summable) with integral $-\infty$. But $\lim\limits_{n \to \infty} f_n(x) = 0$ which obviously has integral 0.

We can avoid this kind of situation and retain the Monotone Convergence Theorem if we keep the f_n bounded below. (A stronger result in this direction is found in Exercise 31.35.)

28.6 Corollary: Let A be a bounded measurable set of real numbers and $\{f_n\}$ a sequence of measurable functions on A such that $M \leqslant f_1 \leqslant f_2 \leqslant \cdots$ for some (finite) constant M. Let $f(x) = \lim\limits_{n \to \infty} f_n(x)$. Then

$$\lim_{n \to \infty} \int_A f_n dm = \int_A f dm.$$

Proof: Use the sequence $0 \leqslant f_1 - M \leqslant f_2 - M \leqslant \cdots$ in the Theorem. Then

$$\lim_{n \to \infty} \int_A (f_n - M) dm = \int_A (\lim_{n \to \infty} f_n - M) dm = \int_A f dm - \int_A M dm.$$

The result follows. □

Of course, if we have a decreasing sequence $\{g_n\}$ of measurable functions bounded above, then the Corollary can be applied to the sequence $\{-g_n\}$ to reach the conclusion $\lim_{n \to \infty} \int_A g_n dm = \int_A g dm$, where $\lim_{n \to \infty} g_n(x) = g(x)$ a.e.

The other major result concerning integration term by term is called the Lebesgue Dominated Convergence Theorem. We will need a Lemma before proving it. Recall that

$$\lim_{n \to \infty} f_n(x) = \lim_{n \to \infty} \{\text{glb} f_k(x) | k \geqslant n\}.$$

(see Appendix).

28.7 Fatou's Lemma: Let A be a bounded measurable set of real numbers. If $\{f_n\}$ is a sequence of non-negative measurable functions on A, and $f(x) = \lim_{n \to \infty} f_n(x)$ for every $x \in A$, then $\int_A f dm \leqslant \lim_{n \to \infty} \int_A f_n dm$.

Proof: Define for each positive integer n and each $x \in A$

$$g_n(x) = \text{glb} \{f_k(x) | k \geqslant n\}.$$

Then each g_n is measurable on A by 18.7 and the g_n's form a monotone increasing sequence of non-negative functions such that $g_n(x) \leqslant f_n(x)$ for each n.

Now $\lim_{n \to \infty} g_n(x) = f(x)$ by definition of $\lim_{n \to \infty} f_n(x)$, so by the Monotone Convergence Theorem $\lim_{n \to \infty} \int_A g_n dm = \int_A f dm$. Since

$$\int_A f_n(x) dm \geqslant \int_A g_n(x) dm$$

for each n, $\lim_{n \to \infty} \int_A f_n dm \geqslant \int_A f dm$. See Exercise 31.10 for this last step. $\square$

28.8 Corollary: Fatou's Lemma holds if $f(x) = \lim_{n \to \infty} f_n(x)$ almost everywhere on A. $\square$

Strict inequality may hold in Fatou's Lemma. See Exercise 31.11 for an example.

28.9 Theorem: (Lebesgue Dominated Convergence Theorem): Let A be a bounded measurable subset of $\mathcal{R}$ and let $\{f_n\}$ be a sequence of measurable functions on A such that $\lim_{n \to \infty} f_n(x) = f(x)$ for every $x \in A$. If there exists a function $g \in \mathcal{L}(A)$ [i.e. g is summable] such that $|f_n(x)| \leqslant g(x)$ for

$n = 1,2,3, \cdots$ and all $x \in A$, then $\lim\limits_{n \to \infty} \int_A f_n dm = \int_A f dm$.

Proof: By Exercise 26.26, f_n and f are in $\mathcal{L}(A)$. Now $f_n + g \geq 0$ on A so by Fatou's Lemma, $\int_A (f + g)dm \leq \lim\limits_{n \to \infty} \int_A (f_n + g)dm$ or $\int_A f dm \leq \lim\limits_{n \to \infty} \int_A f_n dm$ by linearity of the integral (25.4).

It is also true that $g - f_n \geq 0$ on A so by Fatou's Lemma $\int_A (g - f)dm \leq \lim\limits_{n \to \infty} \int_A (g - f_n)dm$ and $-\int_A f dm \leq \lim\limits_{n \to \infty} [-\int_A f_n dm]$.

Thus $\int_A f dm \geq \overline{\lim\limits_{n \to \infty}} \int_A f_n dm$ (see Exercise 31.12). So

$$\int_A f dm \leq \lim\limits_{n \to \infty} \int_A f_n dm \leq \overline{\lim\limits_{n \to \infty}} \int_A f_n dm \leq \int_A f dm.$$

Hence $\lim\limits_{n \to \infty} \int_A f_n dm$ exists and is equal to $\int_A f dm$. $\qquad \square$

28.10 Corollary: The Lebesgue Dominated Convergence Theorem holds when $\lim\limits_{n \to \infty} f_n(x) = f(x)$ and $|f_n(x)| \leq g(x)$ hold almost everywhere on A. $\qquad \square$

28.11 Corollary: If $|f_n(x)| \leq c$ for $n = 1,2,3, \cdots$ and almost all $x \in A$ (i.e. $\{f_n\}$ is a uniformly bounded sequence) and if $f_n \to f$ almost everywhere on A, then $\lim\limits_{n \to \infty} \int_A f_n dm = \int_A f dm$. $\qquad \square$

29.** A Necessary and Sufficient Condition for Riemann Integrability.

Armed with the convergence theorems of the previous section we can determine exactly which functions are Riemann Integrable. This question was far from solved by the existence theorems quoted in section 2. We know that continuous functions and monotone functions are Riemann integrable. Moreover, it is possible to alter a function at finitely many points without affecting integrability. Yet (see Example 3.3) there are Riemann integrable functions which are not even piecewise continuous (continuous at all but finitely many points). It turns out that the discontinuities must be restricted to a set of measure zero. In fact, a bounded function is Riemann integrable if and only if it is continuous almost everywhere.

To prove this result, we need to relate step functions to continuity and to Riemann integrability more intimately than we have heretofor.

29.1 Lemma: A function $f:[a,b] \to \mathcal{R}$ is continuous a.e. if and only if there are sequences $\{g_n\}$ and $\{h_n\}$ of step functions such that

$$g_1 \leqslant g_2 \leqslant \cdots \leqslant f \leqslant \cdots \leqslant h_2 \leqslant h_1$$

and $\lim_{n \to \infty} g_n(x) = f(x) = \lim_{n \to \infty} h_n(x)$ a.e. on $[a,b]$.

Proof: Assume the sequences $\{g_n\}$ and $\{h_n\}$ satisfy the given conditions. The set $\{x \in [a,b] \mid x$ is a point of discontinuity of some g_n or $h_n\}$ is countable, hence has measure 0. If $x' \in [a,b]$ satisfies

$$\lim_{n \to \infty} g_n(x') = f(x') = \lim_{n \to \infty} h_n(x')$$

and x' is not a point of discontinuity of any g_n or h_n (hence for a.e. x'), then given $\epsilon > 0$, there is an n such that $h_n(x') - g_n(x') < \epsilon$. Furthermore, since g_n and h_n are step functions, there is an open interval containing x' in which $h_n(x) = h_n(x')$ and $g_n(x) = g_n(x')$, so that in this interval,

$$g_n(x) - h_n(x') \leqslant f(x) - f(x') \leqslant h_n(x) - g_n(x'),$$

or

$$|f(x) - f(x')| \leqslant |h_n(x') - g_n(x')| < \epsilon.$$

Conversely, let P_n be the regular partition of $[a,b]$ into 2^n equal subintervals. For each subinterval $[y_{i-1}, y_i]$ of P_n, let

$$g_n(x) = \text{glb}\{f(x') \mid x' \in [y_{i-1}, y_i)\}$$

and

$$h_n(x) = \text{lub}\{f(x') \mid x' \in [y_{i-1}, y_i)\}$$

for $x \in [y_{i-1}, y_i)$. Finally, let $g_n(b) = h_n(b) = f(b)$. It is easy to verify that each g_n and h_n is a step function, and since each subinterval of P_n is split into two equal subintervals in P_{n+1}, $g_1 \leqslant g_2 \leqslant \cdots \leqslant f \leqslant \cdots \leqslant h_2 \leqslant h_1$.

If $x' \in [a,b]$ and x' is not in any P_n, and f is continuous at x' (hence for a.e. x'), then for any $\epsilon > 0$, there is a $\delta > 0$ such that if $|x - x'| < \delta$, then $f(x') - \epsilon < f(x) < f(x') + \epsilon$.

If n is so large that some subinterval $[y_{i-1}, y_i]$ of P_n is contained in $(x' - \delta, x' + \delta)$, then

$$g(x') = \text{glb}\{f(x)\,|\,x \in [y_{i-1},y_i)\} \geqslant f(x') - \epsilon,$$

and $h_n(x') \leqslant f(x') + \epsilon$. Therefore, $h_n(x') - g_n(x') \leqslant 2\epsilon$.   ☐

29.2 Theorem: A bounded function $f : [a,b] \to \mathcal{R}$ is Riemann integrable if and only if f is continuous a.e. on $[a,b]$.

Proof: Suppose that f is Riemann integrable. Given n, Theorem 1.6 guarantees the existence of step functions $g_n \leqslant f \leqslant h_n$ such that $\int_a^b h_n(x)dx - \int_a^b g_n(x)dx < 1/n$. Define step functions $\bar{g}_n$ and $\bar{h}_n$ by

$$\bar{g}_n(x) = \max\{g_1(x), \ldots, g_n(x)\}, \qquad \bar{h}_n(x) = \min\{h_1(x), \ldots, h_n(x)\}.$$

Then surely

$$\lim_{n \to \infty} \int_a^b \bar{g}_n(x)dx = \int_a^b f(x)dx = \lim \int_a^b \bar{h}_n(x)dx.$$

If $g(x) = \lim_{n \to \infty} \bar{g}_n(x)$, and $h(x) = \lim_{n \to \infty} \bar{h}_n(x)$, then $g \leqslant f \leqslant h$, and by the Monotone Convergence Theorem,

$$\lim_{n \to \infty} \int_{[a,b]} \bar{g}_n dm = \int_{[a,b]} g\,dm \leqslant \int_{[a,b]} h\,dm = \lim_{n \to \infty} \int_{[a,b]} \bar{h}_n dm.$$

Also, $\int_{[a,b]} \bar{g}_n dm = \int_a^b \bar{g}_n(x)dx$ and $\int_{[a,b]} \bar{h}_n dm = \int_a^b \bar{h}_n(x)dx$, so that

$$\int_a^b f(x)dx = \int_{[a,b]} g\,dm = \int_{[a,b]} h\,dm.$$

Therefore, $g = h$ a.e. (Exercise 26.23) and $f = g = \lim_{n \to \infty} \bar{g}_n$ a.e. Thus f is measurable and f is continuous a.e. by Lemma 29.1.

For the converse, suppose f is continuous almost everywhere. Let g_n, h_n be as in Lemma 29.1. Then by the Monotone Convergence Theorem (in particular, Corollary 28.6 and the discussion following it),

$$\lim_{n \to \infty} \int_{[a,b]} g_n dm = \lim_{n \to \infty} \int_{[a,b]} h_n dm.$$

Hence, given $\epsilon > 0$, there exists an n such that

$$\int_a^b h_n(x)dx - \int_a^b g_n(x)dx = \int_{[a,b]} h_n dm - \int_{[a,b]} g_n dm < \epsilon.$$

Riemann integrability follows from Theorem 1.6.   ☐

29.3 Corollary: If f is Riemann integrable on $[a,b]$, then f is Lebesgue integrable on $[a,b]$, and $\int_a^b f(x)dx = \int_{[a,b]} f dm$.

Proof: In the proof of the theorem, we showed that f is measurable. (it is also measurable by virtue of being continuous a.e.). Since f is bounded, f is Lebesgue integrable. Since $f = g$ a.e.,

$$\int_{[a,b]} f dm = \int_{[a,b]} g dm = \lim_{n \to \infty} \int_{[a,b]} \bar{g}_n dm = \lim_{n \to \infty} \int_a^b \bar{g}_n(x)dx = \int_a^b f(x)dx.$$
$\square$

The meaning of this corollary is that all the functions you could integrate in calculus can be integrated in the Lebesgue sense as well, using the same techniques. Furthermore, the Lebesgue values are the same as the Riemann values.

30.** Egoroff's and Lusin's Theorems and an Alternative Proof of the Lebesgue Dominated Convergence Theorem.

In this section we will prove two interesting and useful theorems. The first, Egoroff's Theorem, says that a sequence $\{f_n\}$ of measurable functions which converges to a function f almost everywhere on a bounded measurable set A is somehow close to being a uniformly convergent sequence. "Close to" means that for any $\epsilon > 0$ there is a measurable set $A_\epsilon \subset A$ such that $m(A \backslash A_\epsilon) < \epsilon$ and $f_n \to f$ uniformly on A_ϵ. One can imagine some uses of such a result. In particular, convergence theorems involving the Lebesgue integral may be proved using Egoroff's Theorem since by Exercise 31.1 uniform convergence allows integration term by term.

30.1 Theorem (Egoroff): Let A be a bounded measurable set in $\mathcal{R}$. Let f_n and f be functions defined on A such that each f_n is measurable and $f_n \to f$ almost everywhere on A. Then given any $\epsilon > 0$, there exists a measurable subset $A_\epsilon \subset A$ such that $m(A \backslash A_\epsilon) < \epsilon$ and $f_n \to f$ uniformly on A_ϵ.

Proof: Given $\epsilon > 0$ and positive integers m and n, define the set $E_{m,n} = \bigcap_{i=n}^\infty \{x \mid |f_i(x) - f(x)| < 1/m\}$. Note that $E_{m,n}$ is measurable. If U is the subset on which $f_n \to f$, then clearly for any m, $U \subset \bigcup_{n=1}^\infty E_{m,n} \subset A$.

Now $m(U) = m(A)$ so $m(\bigcup_{n=1}^\infty E_{m,n}) = m(A)$. But $E_{m,n} \subset E_{m,n+1}$ for all m and n, so

$$\lim_{n \to \infty} m(A \setminus E_{m,n}) = \lim_{n \to \infty} [m(A) - m(E_{m,n})]$$

$$= m(A) - \lim_{n \to \infty} m(E_{m,n}) = m(A) - m(\bigcup_{n=1}^{\infty} E_{m,n}) = 0.$$

Thus for each m, there exists an integer n_m such that

$$m(A \setminus E_{m,n_m}) < \epsilon/2^m.$$

Now let $A_\epsilon = \bigcap_{m=1}^{\infty} E_{m,n_m}$. Then A_ϵ is measurable and

$$m(A \setminus A_\epsilon) = m(A \setminus \bigcap_{m=1}^{\infty} E_{m,n_m}) = m(\bigcup_{m=1}^{\infty} (A \setminus E_{m,n_m}))$$

$$\underset{8.6}{\leqslant} \sum_{m=1}^{\infty} m(A - E_{m,n_m}) < \sum_{m=1}^{\infty} \epsilon/2^m = \epsilon.$$

We now claim that $f_n \to f$ uniformly on A_ϵ since given any m, there exists an n_m such that for all $n > n_m$, $|f_n(x) - f(x)| < 1/m$ everywhere on E_{m,n_m}. But $A_\epsilon \subset E_{m,n_m}$ for every m, so for all $n > n_m$, $|f_n(x) - f(x)| < 1/m$ everywhere on A_ϵ. But this is exactly uniform convergence. $\square$

We will now use Egoroff's Theorem to give another proof of the Lebesgue Dominated Convergence Theorem.

30.2 Corollary (*Lebesgue Dominated Convergence Theorem*): Given measurable functions f_n converging to f on a bounded measurable set A with $|f_n(x)| < g(x)$ for some $g \in \mathcal{L}(A)$, then $\lim_{n \to \infty} \int_A f_n dm = \int_A f dm$.

Proof: Clearly f is also measurable and $|f| \leqslant g$ implies $f \in \mathcal{L}(A)$. Define disjoint measurable sets $A_k = \{x | k - 1 \leqslant g(x) < k\}$ for $k = 1, 2, \cdots$. Then $A = \bigcup_{k=1}^{\infty} A_k$ so that $\int_A g dm = \sum_{k=1}^{\infty} \int_{A_k} g dm$ by Theorem 24.3. Thus given $\epsilon > 0$, there is an N such that

$$\int_{\substack{\bigcup A_k \\ k=N+1}}^{\infty} g dm = \sum_{k=N+1}^{\infty} \int_{A_k} g dm < \epsilon/5.$$

Thus $\int_{\substack{\bigcup A_k \\ k=N+1}}^{\infty} |f_n| dm$ and $\int_{\substack{\bigcup A_k \\ k=N+1}}^{\infty} |f| dm$ are each $< \epsilon/5$. This takes care of the set on which f_n and f are large. On $\bigcup_{k=1}^{N} A_k$ use Egoroff's Theorem to write $\bigcup_{k=1}^{N} A_k$ as $B_1 \cup B_2$ where $m(B_1) < \epsilon/5N$ and

$$\int_{B_2} |f_n - f| dm < \epsilon/5m(A)$$

for large enough n (by uniform convergence).

Now for large enough n:

$$\left| \int_A f_n dm - \int_A f dm \right| \leq \int_{\substack{\cup A_k \\ k=N+1}}^{\infty} |f_n| dm + \int_{\substack{\cup A_k \\ k=N+1}}^{\infty} |f| dm$$

$$+ \int_{B_1} |f_n| dm + \int_{B_1} |f| dm + \int_{B_2} |f_n - f| dm < \epsilon.$$

□

This alternative proof of the *Dominated Convergence Theorem* shows the power of *Egoroff's Theorem*. The proof is in many ways more intuitive and natural than the one in section 29, but it requires the uniform convergence provided by Egoroff's Theorem.

We now proceed to Lusin's Theorem which says that a measurable function f on a bounded measurable set A is "nearly" continuous. "Nearly" here means given any $\epsilon > 0$, there is a subset C of A such that $m(A \setminus C) < \epsilon$ and f is continuous on C.

30.3 Theorem (*Lusin*): If f is a measurable function on a bounded measurable set A, then given any $\epsilon > 0$, there exists a closed set $C_\epsilon \subset A$ such that $m(A \setminus C_\epsilon) < \epsilon$ and f is continuous on C_ϵ. (That is, the restriction of f to C_ϵ is continuous.)

Proof: First, let $f = \sum_{k=1}^{n} a_k \chi_{E_k}$ be a simple function on A. Given $\epsilon > 0$, for each k there exists a closed set $C_k \subset E_k$ such that

$$m(E_k \setminus C_k) < \epsilon/n$$

(see Corollary 13.3). Now $C = \bigcup_{k=1}^{n} C_k$ is closed, $m(A \setminus C) < \epsilon$ (Exercise 31.23) and f is continuous on C since C is equal to the disjoint union of closed sets C_k and f is constant on each C_k. (Exercise 31.24).

Now let f be any measurable function. Since $f = f_+ - f_-$, we may assume f is non-negative. Since f is measurable, $f = \lim_{n \to \infty} f_n$ for a sequence $\{f_n\}$ of simple functions by Theorem 19.4.

By the first part of the proof, given $\epsilon > 0$, there exists for each n a closed set C_n such that $m(A \setminus C_n) < \epsilon/2^{n+1}$ and f_n is continuous on C_n.

Let $C_0 = \bigcap_{n=1}^{\infty} C_n$. Then C_0 is closed and

$$m(A\backslash C_0) = m(A \backslash \bigcap_{n=1}^{\infty} C_n) = m(\bigcup_{n=1}^{\infty} (A\backslash C_n))$$

$$\underset{8.6}{\leqslant} \sum_{n=1}^{\infty} m(A\backslash C_n) < \epsilon \sum_{n=1}^{\infty} 1/2^{n+1} = \epsilon/2.$$

We would like to conclude that f is continuous on C_0, but we cannot since we need something like uniform convergence for this. However, by Egoroff's Theorem, there is a measurable subset $C \subset C_0$ such that $m(C_0\backslash C) < \epsilon/4$ [and thus $m(A\backslash C) = m(A\backslash C_0) + m(C_0\backslash C) < \frac{3}{4}\epsilon]$, and $f_n \to f$ uniformly on C.

Now each f_n is continuous on $C \subset C_0$ so that f, being the uniform limit of the f_n, is continuous on C. If C is closed we are through. If not, take closed set $F \subset C$ such that $m(C\backslash F) < \epsilon/4$. □

31. Exercises

31.1 Show that if $f_n \to f$ uniformly on a bounded measurable set A, and each $f_n \in \mathcal{L}(A)$, then $f \in \mathcal{L}(A)$, and $\lim_{n\to\infty} \int_A f_n dm = \int_A f dm$.

31.2 Consider the sequence of functions $f_n(x) = n^2 x(1 - x^2)^n$, $x \in [0,1]$.
  (a) Show that $\lim_{n\to\infty} f_n(x) = 0$ for all $x \in [0,1]$.
  (b) Show that $\lim_{n\to\infty} \int_{[0,1]} f_n dm = \infty$.
  (c) What happens to the sequence $g_n(x) = nx(1 - x^2)^n$, $x \in [0,1]$?

31.3 Prove Lemma 28.1. (Hint: use countable additivity of the integral (Theorem 25.5) to show that $\int_{\cup A_n} g dm = \lim_{n\to\infty} \int_{A_n} g dm$.)

31.4 Prove Corollary 28.3 in detail.

31.5 Use the Monotone Convergence Theorem to show that $f(x) = 1/x$ is not summable on $(0,1)$. (Hint: consider the sequence $f_n(x) = \max\{f(x),n\}$, $x \in (0,1)$.)

31.6 Use the Monotone Convergence Theorem to find $\int_{(0,1)} g dm$, where $g(x) = x^{-\frac{1}{2}}$.

31.7 Suppose that h is a continuous function on $[a,b)$, but $\lim_{x\to b^-} h(x) = +\infty$. Then of course the *improper* Riemann integral $\int_a^b h(x)dx$ is defined by

$\lim_{t \to b} \int_a^t h(x)dx$. Show that if this limit exists and is finite, then h is Lebesgue integrable on $[a,b)$, and $\int_{[a,b)} h\, dm = \int_a^b h(x)dx$.

31.8 Give an example to show that monotonicity in the Monotone Convergence Theorem is essential.

31.9 Let $f \in \mathcal{L}[0,1]$ and $g_n(x) = x^n f(x)$ for $x \in [0,1]$, $n = 1,2,\ldots$. Show that $g_n \in \mathcal{L}[0,1]$ for $n = 1,2,\ldots$, and $\lim_{n \to \infty} \int_{[0,1]} g_n\, dm = 0$.

31.10 Suppose that $a_n \geqslant b_n$ for all n. Show that $\lim_{n \to \infty} a_n \geqslant \lim_{n \to \infty} b_n$.

31.11 Show that strict inequality may hold in the conclusion of Fatou's Lemma.

31.12 If $-\alpha \leqslant \lim_{n \to \infty}(-a_n)$, show that $\alpha \geqslant \overline{\lim}_{n \to \infty} a_n$.

31.13 Where is the fact that g is *summable* used in the proof of the Dominated Convergence Theorem?

31.14 Let $f_n \geqslant 0, f_n \in \mathcal{L}(A), f_n \to 0$, a.e. on A. Define

$$g_n(x) = \max\{f_1(x), \ldots, f_n(x)\}$$

for $x \in A$, and assume that $\int_A g_n\, dm \leqslant M$ for some real number M, all n. Prove that $\lim_{n \to \infty} \int_A f_n\, dm = 0$.

31.15 Let $f_n \to f$, $|f_n| \leqslant h$, $g,h \in \mathcal{L}(A)$. Show that

$$\lim_{n \to \infty} \int_A f_n g\, dm = \int_A fg\, dm.$$

31.16 Give an example of an *unbounded* function (hence non-Riemann integrable) which is continuous a.e. on $[0,1]$.

31.17 Give an easy proof that the function g of Example 3.3 is Riemann integrable.

31.18 Show that χ_C, where C is the Cantor set (Example 12.5 and Exercises 16.20 and 16.21) is Riemann integrable, but that χ_D (Exercise 16.23) is not.

31.19 If $A \subset [a,b]$ is closed, show that χ_A is Riemann integrable on $[a,b]$ if $m(A) = 0$.

31.20 Let $f,h \in \mathcal{L}(A)$ and $\int_A f\, dm = \int_A h\, dm$. If $f \leqslant g \leqslant h$ on A, show that $g \in \mathcal{L}(A)$ and $\int_A f\, dm = \int_A g\, dm$.

31.21 Give an example to show that Egoroff's Theorem cannot be improved to yield $m(A_\epsilon) = 0$.

31.22 For each example in this chapter of an a.e. pointwise convergent sequence of functions, find an A_ϵ as in Egoroff's Theorem.

31.23 Let $E_1, \ldots, E_n$ be disjoint measurable bounded sets. Let $C_k \subset E_k$ be measurable for $k = 1, \ldots, n$. Show that

$$m(\bigcup_{k=1}^{n} E_k \setminus \bigcup_{k=1}^{n} C_k) = \sum_{k=1}^{n} m(E_k \setminus C_k).$$

31.24 Prove that if $C_1, \ldots, C_n$ are disjoint closed sets and f is constant on each C_k, then f is continuous on $\bigcup_{k=1}^{n} C_k$. Is this true under any other assumptions about the C_k? (Hint: you must show that for every $x \in C_k$, there is a $\delta > 0$ such that $(x - \delta, x + \delta) \cap C_j = \emptyset$ for $j \neq k$. Why does this prove the result?)

31.25 Let $f = \chi_Q$ on $[0,1]$. Given $\epsilon > 0$, find C_ϵ as in Lusin's Theorem.

31.26 Prove Lusin's Theorem for f non-negative, measurable, and *bounded* without using Egoroff's Theorem. (Hint: see Corollary 19.5.)

31.27 If $f: C \to \mathcal{R}$ is continuous on the closed set C, show that f can be extended to a continuous function on all of $\mathcal{R}$. (Hint: $\mathcal{R} \setminus C = \bigcup_i I_i$, where the I_i are disjoint open intervals.)

31.28 Use Exercise 31.27 to prove the following corollary of Lusin's Theorem: if f is measurable on a bounded measurable set A, and $\epsilon > 0$, then there is a continuous function g such that $m(\{x \in A \mid f(x) \neq g(x)\}) < \epsilon$.

31.29 Prove the following converse of Exercise 31.28: Suppose that A is a bounded measurable set and $f: A \to \mathcal{R}$. If for every $\epsilon > 0$ there is a continuous function g such that $m(\{x \in A \mid f(x) \neq g(x)\}) < \epsilon$, then f is measurable on A. (Hint: by taking $\epsilon = 1, 1/2, 1/3, \ldots$, obtain a sequence $\{g_n\}$ of continuous functions which converge pointwise a.e. to f.)

31.30 Show that if f is a measurable function on A, then f is the a.e. pointwise limit of a sequence of continuous functions. (Hint: see Exercise 31.29.)

31.31 Discuss possible definitions of $\int_A f \, dm$ for A unbounded and measurable, f measurable on A. Restrict yourself to $A = \mathcal{R}$ if you wish. What theorems follow from your definition?

31.32 Find $\int_{[0,1]} f \, dm$, where (a) $f(x) = x$, and (b) $f(x) = x^3 + 2x$.

31.33 Let $\{h_n\}$ be an increasing sequence of functions in $\mathcal{L}(A)$, with $h_n \leqslant h$ for all n and some $h \in \mathcal{L}(A)$. If $\lim_{n \to \infty} \int_A h_n \, dm = \int_A h \, dm$, prove that

$h = \lim\limits_{n \to \infty} h_n$ a.e. on A. (Hint: As in the proof of Theorem 29.2, let $g(x) = \lim\limits_{n \to \infty} h_n(x)$, and show that $g = h$ a.e. in A.)

31.34 Let A be bounded and measurable, $f: A \to \mathfrak{R}$, and suppose that for all $\epsilon > 0$, there are simple functions $g \leqslant f \leqslant h$ such that $\int_A h\, dm - \int_A g\, dm < \epsilon$. Show that f is bounded and measurable. (See the proof of Theorem 29.2).

31.35 Let A be bounded and measurable and let $\{f_n\}$ be a sequence of measurable functions on A such that $f_1 \leqslant f_2 \leqslant \ldots$, where $\int_A f_1\, dm$ is finite. Prove that if $f(x) = \lim\limits_{n \to \infty} f_n(x)$ for $x \in A$, then $\lim\limits_{n \to \infty} \int_A f_n\, dm = \int_A f\, dm$.

CHAPTER 7

Function Spaces and L^2

32. Linear Spaces

Having developed the main points of the Lebesgue integration theory, we will devote the remainder of the book to theoretical applications in analysis. We will touch on only a few of the uses to which the theory has been put, but hopefully these will give some feeling for how important the Lebesgue integral has become in modern analysis.

One of the most fruitful recent developments in analysis has been to apply the techniques and concepts of linear algebra to the study of functions. In particular, certain sets of functions have natural structures as vector spaces (in this context, they are called *function* spaces), and their vector space properties yield a great deal of information about the functions themselves.

We assume the reader has some familiarity with the basic notions of linear algebra. However, in this section we will review some of the definitions and results which will be most useful to us in the remainder of the book.

32.1 Definition: A real *vector space* (or *linear space*) is a non-empty set V with two operations. The first assigns to each $\bar{v}, \bar{w} \in V$ a unique element $\bar{v} + \bar{w} \in V$. The second assigns to each $a \in \mathcal{R}$ and $\bar{v} \in V$, a unique element $a\bar{v} \in V$. These operations must satisfy the following properties:

(1) $\bar{u} + \bar{v} = \bar{v} + \bar{u}$ for all $\bar{u}, \bar{v} \in V$;

(2) $\bar{u} + (\bar{v} + \bar{w}) = (\bar{u} + \bar{v}) + \bar{w}$ for all $\bar{u}, \bar{v}, \bar{w} \in V$;

(3) there is an element $\bar{0} \in V$ such that $\bar{u} + \bar{0} = \bar{u}$ for all $\bar{u} \in V$;

(4) for each $\bar{u} \in V$, there is an element $-\bar{u} \in V$ such that $\bar{u} + (-\bar{u}) = \bar{0}$;

(5) $a(\bar{u} + \bar{v}) = a\bar{u} + a\bar{v}$ for all $a \in R, \bar{u}, \bar{v} \in V$;

(6) $(a + b)\bar{u} = a\bar{u} + b\bar{u}$ for all $a, b \in R, \bar{u} \in V$;

(7) $(ab)\bar{u} = a(b\bar{u})$ for all $a, b \in R, \bar{u} \in V$;

(8) $1\bar{u} = \bar{u}$ for all $\bar{u} \in V$.

32.2 Examples: First, the prime example.

(1) The set of all n-tuples of real numbers, denoted R^n, with addition and scalar multiplication defined respectively by

$$(x_1, x_2, \ldots, x_n) + (y_1, y_2, \ldots, y_n) = (x_1 + y_1, x_2 + y_2, \ldots, x_n + y_n),$$

$$a(x_1, x_2, \ldots, x_n) = (ax_1, ax_2, \ldots, ax_n).$$

The following examples are all function spaces. The operations are defined by

$$(f + g)(x) = f(x) + g(x),$$

and

$$(af)(x) = af(x).$$

Most of the vector space properties are trivial to verify in each example below. The only remaining requirement in each case is to show that if f and g are in the set in question, so are $f + g$ and af. In each case this is either trivial or an immediate consequence of elementary facts proved in calculus or in this book.

(2) $B(A) = \{g \,|\, g : A \to R, g \text{ bounded}\}$, for any $A \subset R$.

(3) $C(A) = \{g \,|\, g : A \to R, g \text{ continuous}\}$, any $A \subset R$.

(4) $R([a,b]) = \{g \,|\, g : [a,b] \to R, g \text{ Riemann integrable}\}$.

(5) $L(A) = \{g \,|\, g : A \to R, g \text{ Lebesgue summable}\}$, A bounded and measurable.

(6) $P(A) = \{g \,|\, g : A \to R, g \text{ a polynomial}\}$.

(7) $P_n(A) = \{g \,|\, g \in P, g \text{ of degree} \leqslant n\}$.

The concept of *length* of a vector is fundamental because it leads immediately to a definition of *distance* between two vectors. Distance is particularly important in analysis, of course, since so much of analysis is concerned with approximations (as in limits, integrals, etc.). Based on the connection of $\mathcal{R}^3$ with Euclidean geometry, length in $\mathcal{R}^n$ is most reasonably taken to be Pythagorean length

$$(\text{length of } (x_1, \ldots, x_n)) = \sqrt{x_1^2 + \ldots + x_n^2}).$$

In the function spaces, however, it is much less clear how best to define this concept.

32.3 Definition: A real *normed* linear space is a vector space V (with addition and scalar multiplication denoted by $\bar{u} + \bar{v}, a\bar{u}$, respectively) such that for each $\bar{u} \in V$, there is a unique real number $\|\bar{u}\|$ (called the *norm* or *length* of $\bar{u}$) satisfying:

(1) $\|\bar{u}\| \geqslant 0$ for all $\bar{u} \in V$;
(2) $\|\bar{u}\| = 0$ if and only if $\bar{u} = \bar{0}$;
(3) $\|a\bar{u}\| = |a| \|\bar{u}\|$ for all $a \in \mathcal{R}, \bar{u} \in V$;
(4) $\|\bar{u} + \bar{v}\| \leqslant \|\bar{u}\| + \|\bar{v}\|$ for all $\bar{u}, \bar{v} \in V$.

32.4 Examples:

(1) For $(x_1, \ldots, x_n) \in \mathcal{R}^n$, $\|(x_1, \ldots, x_n)\| = \sqrt{x_1^2 + x_2^2 + \ldots + x_n^2}$.
(2) For $f \in \mathcal{B}(A)$, $\|f\| = \text{lub}\{|f(x)| \mid x \in A\}$.
(3) For $f \in C([0,1])$, $\|f\| = \int_0^1 |f(x)| dx$.
(4) For $f \in P_1([0,1])$, where $f(x) = a_0 + a_1 x$,

$$\|f\| = \sqrt{\int_0^1 f^2(x) dx} = \sqrt{a_0^2 + a_0 a_1 + \tfrac{1}{3} a_1^2}.$$

Whenever a space has a norm, then a distance, or *metric* can be immediately defined by $d(\bar{u}, \bar{v}) = \|\bar{u} - \bar{v}\|$. This has the properties, for all $\bar{u}, \bar{v}, \bar{w} \in V$,

(1) $d(\bar{u}, \bar{v}) \geqslant 0$;
(2) $d(\bar{u}, \bar{v}) = 0$ if and only if $\bar{u} = \bar{v}$;
(3) $d(\bar{u}, \bar{v}) = d(\bar{v}, \bar{u})$;
(4) $d(\bar{u}, \bar{w}) \leqslant d(\bar{u}, \bar{v}) + d(\bar{v}, \bar{w})$.

So, a normed linear space is a metric space, and one can talk about limits, continuity, etc., for real valued functions defined on the space. We are most concerned with limits of sequences.

32.13 Cauchy Inequality: If V is an inner product space with inner product $\overline{u} \cdot \overline{v}$, then for any $\overline{u}, \overline{v} \in V$,

$$|\overline{u} \cdot \overline{v}| \leqslant \sqrt{\overline{u} \cdot \overline{u}} \sqrt{\overline{v} \cdot \overline{v}}.$$

Proof: If $\overline{u} = \overline{0}$ or $\overline{v} = \overline{0}$, then equality holds trivially. Otherwise, consider the vectors

$$\overline{u}' = \frac{\overline{u}}{\sqrt{\overline{u} \cdot \overline{u}}} \text{ and } \overline{v}' = \frac{\overline{v}}{\sqrt{\overline{v} \cdot \overline{v}}}.$$

It is easy to verify that $\overline{u}' \cdot \overline{u}' = \overline{v}' \cdot \overline{v}' = 1$, so that by the lemma, $|\overline{u}' \cdot \overline{v}'| \leqslant 1$. But

$$|\overline{u}' \cdot \overline{v}'| = \frac{|\overline{u} \cdot \overline{v}|}{\sqrt{\overline{u} \cdot \overline{u}} \sqrt{\overline{v} \cdot \overline{v}}} \leqslant 1. \qquad \square$$

32.14 Corollary: In an inner product space V, if $\|\overline{u}\| = \sqrt{\overline{u} \cdot \overline{u}}$, then $\| \ \|$ is a norm.

Proof: The defining properties (1) - (3) of Definition 32.3 are left to the reader (Exercise 34.17). For property (4), we calculate:

$$\|\overline{u} + \overline{v}\|^2 = (\overline{u} + \overline{v}) \cdot (\overline{u} + \overline{v})$$

$$= \overline{u} \cdot \overline{u} + 2(\overline{u} \cdot \overline{v}) + \overline{v} \cdot \overline{v}$$

$$= \|\overline{u}\|^2 + 2(\overline{u} \cdot \overline{v}) + \|\overline{v}\|^2$$

$$\underset{32.13}{\leqslant} \|\overline{u}\|^2 + 2\|\overline{u}\| \|\overline{v}\| + \|\overline{v}\|^2$$

$$= (\|\overline{u}\| + \|\overline{v}\|)^2. \qquad \square$$

32.15 Examples: (1) The norm of Example 32.4(1) comes from the dot product:

$$\|(x_1, \ldots, x_n)\| = \sqrt{x_1^2 + \ldots + x_n^2} = \sqrt{(x_1, \ldots, x_n) \cdot (x_1, \ldots, x_n)}.$$

(2) The norm on $P_1([0,1])$ defined by

$$\|a_0 + a_1 x\| = \sqrt{a_0^2 + a_0 a_1 + \frac{1}{3}a_1^2}$$

(Example 32.4(4)) comes from the inner product $f \cdot g = \int_0^1 f(x)g(x)dx$.

(3) The norm on $C([0,1])$ defined by $\|f\| = \int_0^1 |f(x)|dx$ (Example 32.4(3)) does not come from any inner product. Indeed, if we supposed there were an inner product such that $f \cdot f = (\int_0^1 |f(x)|dx)^2$, then we would have

$$1 \cdot 1 = \left(\int_0^1 1\,dx\right)^2 = 1, \qquad x \cdot x = \left(\int_0^1 x\,dx\right)^2 = \frac{1}{4},$$

$$(1 + x) \cdot (1 + x) = \left(\int_0^1 (1 + x)dx\right)^2 = \left(1 + \frac{1}{2}\right)^2 = 9/4.$$

But $(1 + x) \cdot (1 + x) = 1 \cdot 1 + x \cdot x + 2(1 \cdot x) = 1 + \frac{1}{4} + 2(1 \cdot x)$. Hence $1 \cdot x = \frac{1}{2}$. Then $(\frac{1}{2} - x) \cdot (\frac{1}{2} - x) = \frac{1}{2} \cdot \frac{1}{2} - 2(\frac{1}{2} \cdot x) + x \cdot x = \frac{1}{4} - \frac{1}{2} + \frac{1}{4} = 0$, even though $\frac{1}{2} - x$ is not the zero function (contradicting Definition 32.10(2)).

An inner product space that is complete in its norm is called a *Hilbert* space.

The inner product is useful for defining angles in our space, in particular right angles.

32.16 Definition: If V is an inner product space with inner product $u \cdot v$, then the set $\{\bar{u}_1, \bar{u}_2, \ldots\} \subset V$ is an *orthogonal* set if for every $i \neq j$, $\bar{u}_i \cdot \bar{u}_j = 0$. The set is *orthonormal* if it is orthogonal and for each i, $\bar{u}_i \cdot \bar{u}_i = 1$, that is $\|\bar{u}_i\| = 1$.

32.17 Example: (1) In R^3, the set $\{(1,0,0),(0,1,0),(0,0,1)\}$ is orthonormal relative to the dot product.

(2) In $C([0,1])$, the set consisting of $f_1(x) = 1, f_2(x) = x - 1/2$, $f_3(x) = x^2 - x + 1/6, f_4(x) = x^3 - \frac{3}{2}x^2 + \frac{3}{5}x - \frac{1}{20}$ is orthogonal relative to the inner product $f \cdot g = \int_0^1 f(x)g(x)dx$.

(3) In $C([-\pi,\pi])$, the set

$$\left\{ \frac{1}{\sqrt{2\pi}}, \frac{1}{\sqrt{\pi}}\sin x, \frac{1}{\sqrt{\pi}}\cos x, \frac{1}{\sqrt{\pi}}\sin 2x, \frac{1}{\sqrt{\pi}}\cos 2x, \ldots \right\}$$

is orthonormal relative to the inner product $f \cdot g = \int_{[-\pi,\pi]} fg\,dm$. (See Exercise 34.20.)

You will recall from linear algebra that orthogonal sets are linearly independent (see any linear algebra book). Furthermore, if $\{\bar{v}_1, \bar{v}_2, \ldots, \bar{v}_n\}$ is an orthonormal basis of V, then any element $\bar{v} \in V$ can be expressed as $\bar{v} = \sum_{i=1}^{n} (\bar{v} \cdot \bar{v}_i)\bar{v}_i$ (see Exercise 34.21). The following is an important related idea which we will find useful later.

Let $\{\bar{v}_1, \ldots, \bar{v}_n\}$ be an orthonormal set (not necessarily a basis) in an inner product space V. Consider the subspace W spanned by $\{\bar{v}_1, \ldots, \bar{v}_n\}$ (i.e., W is the set of all linear combinations $a_1\bar{v}_1 + \ldots + a_n\bar{v}_n$, where $a_1, \ldots, a_n \in \mathcal{R}$). Then, given $\bar{v} \in V$, the vector

$$\bar{v}_W = (\bar{v} \cdot \bar{v}_1)\bar{v}_1 + \ldots + (\bar{v} \cdot \bar{v}_n)\bar{v}_n$$

is an element of W, and is called the *projection of $\bar{v}$ on W*. Note that if $\bar{v} \in W$, then $\bar{v}_W = \bar{v}$.

32.18 Theorem: Given $\bar{v} \in V$, $\bar{v}_W$ is the vector in W which is closest to $\bar{v}$ (in the sense of the norm). That is, for any $\bar{w} \in W$, $\|\bar{v} - \bar{w}\| \geqslant \|\bar{v} - \bar{v}_W\|$.

Proof: Let $\bar{w} = \sum_{i=1}^{n} a_i\bar{v}_i$ be an arbitrary element of W. Then

$$\|\bar{v} - \bar{w}\|^2 = (\bar{v} - \sum_{i=1}^{n} a_i\bar{v}_i) \cdot (\bar{v} - \sum_{j=1}^{n} a_j\bar{v}_j)$$

$$= \bar{v} \cdot \bar{v} - \sum_{i=1}^{n} a_i(\bar{v} \cdot \bar{v}_i) - \sum_{j=1}^{n} a_j(\bar{v} \cdot \bar{v}_j) + \sum_{i=1}^{n} \sum_{j=1}^{n} a_i a_j(\bar{v}_i \cdot \bar{v}_j)$$

$$= \|\bar{v}\|^2 - 2\sum_{i=1}^{n} a_i(\bar{v} \cdot \bar{v}_i) + \sum_{i=1}^{n} a_i^2,$$

since

$$\bar{v}_i \cdot \bar{v}_j = \begin{cases} 1 \text{ if } i = j \\ 0 \text{ if } i \neq j \end{cases}$$

and

$$\sum_{i=1}^{n} a_i(\bar{v} \cdot \bar{v}_i) = \sum_{j=1}^{n} a_j(\bar{v} \cdot \bar{v}_j).$$

A similar calculation gives

$$\| \bar{v} - \bar{v}_W \|^2 = \| \bar{v} - \sum_{i=1}^{n} (\bar{v} \cdot \bar{v}_i) \bar{v}_i \|^2$$

$$= \| \bar{v} \|^2 - 2 \sum_{i=1}^{n} (\bar{v} \cdot \bar{v}_i)(\bar{v} \cdot \bar{v}_i) + \sum_{i=1}^{n} (\bar{v} \cdot \bar{v}_i)^2$$

$$= \| \bar{v} \|^2 - \sum_{i=1}^{n} (\bar{v} \cdot \bar{v}_i)^2.$$

Therefore

$$\| \bar{v} - \bar{w} \|^2 - \| \bar{v} - \bar{v}_W \|^2 = \sum_{i=1}^{n} a_i^2 - 2 \sum_{i=1}^{n} a_i (\bar{v} \cdot \bar{v}_i) + \sum_{i=1}^{n} (\bar{v} \cdot \bar{v}_i)^2$$

$$= \sum_{i=1}^{n} (a_i^2 - 2a_i (\bar{v} \cdot \bar{v}_i) + (\bar{v} \cdot \bar{v}_i)^2)$$

$$= \sum_{i=1}^{n} (a_i - (\bar{v} \cdot \bar{v}_i))^2 \geqslant 0. \qquad \square$$

32.19 Corollary: Under the hypotheses of the theorem,

$$\| v - \sum_{i=1}^{n} (\bar{v} \cdot \bar{v}_i) \bar{v}_i \|^2 = \| \bar{v} \|^2 - \sum_{i=1}^{n} (\bar{v} \cdot \bar{v}_i)^2,$$

and

$$\sum_{i=1}^{n} (\bar{v} \cdot \bar{v}_i)^2 \leqslant \| \bar{v} \|^2.$$

Proof: From the proof of the theorem,

$$\| \bar{v} \|^2 - \sum_{i=1}^{n} (\bar{v} \cdot \bar{v}_i)^2 = \| \bar{v} - \sum_{i=1}^{n} (\bar{v} \cdot \bar{v}_i) \bar{v}_i \|^2 \geqslant 0. \qquad \square$$

33. The Space $\mathcal{L}^2$

Given a bounded measurable set A, we wish to define a particularly useful norm for certain functions $f : A \to \mathcal{R}$. This norm is a generalization of the Pythagorean norm in $\mathcal{R}^n$ given by $\| (y_1, \ldots, y_n) \| = \sqrt{y_1^2 + \ldots + y_n^2}$. Now we could represent the n-tuple $(y_1, \ldots, y_n)$ by a step function

$$f : [0, n) \to \mathcal{R}$$

defined by

$$f(x) = y_i \text{ for } x \in [i-1, i).$$

Then

$$\|(y_1, \ldots, y_n)\| = \sqrt{\int_{[0,n)} f^2 dm}.$$

For a general function $f: A \to \mathcal{R}$, we attempt to define by analogy the $\mathcal{L}^2$ norm

$$\|f\|_2 = \sqrt{\int_A f^2 dm}.$$

Of course, this will not be defined for every function; since the norm should be finite for every f, we need $f^2 \in \mathcal{L}(A)$. Hence f^2 must be measurable. This does not imply f is measurable (see Exercise 34.24), but for technical reasons we will require f to be measurable as well.

33.1 Definition: If A is bounded and measurable, let

$$\mathcal{L}^2(A) = \left\{ f: A \to \mathcal{R} \,\middle|\, f \text{ is measurable and } \int_A f^2 dm < \infty \right\}.$$

That is, $\mathcal{L}^2(A)$ is the set of measurable, square summable functions on A. For $f \in \mathcal{L}^2(A)$, define $\|f\|_2 = \sqrt{\int_A f^2 dm}$.

We will verify that $\| \ \|_2$ satisfies properties (1), (3), and (4) of Definition 32.3, so that it is almost a norm. Unfortunately (2) is false, since $\int_A f^2 dm = 0$ only implies that $f = 0$ a.e. on A. The way out of this dilemma is to replace our notion of "equality" with the modified notion of "equality almost everywhere." Therefore, we will consider two functions $f, g \in \mathcal{L}^2(A)$ to be identical if and only if $f = g$ a.e. on A. Of course, this is a somewhat bizarre and informal procedure. It could be made rigorous by defining $\mathcal{L}^2(A)$ to be the set of equivalence classes—under the equivalence relation" = a.e." (see Exercise 34.25)—of measurable square summable functions on A. This procedure has the virtue of being logically unobjectionable, but involves a notational nightmare. We will therefore follow the universal practice of identifying functions which are equal almost everywhere. Thus property (2) of Definition 32.3 holds for $\| \ \|_2$. The other properties will follow from the discussion below.

First, we should verify that $\mathcal{L}^2(A)$ is a linear space (that is, a function space) under the ordinary operations on functions.

33.2 Proposition: $\mathcal{L}^2(A)$ is a real linear space.

Proof: By the properties of addition and multiplication of real numbers, in terms of which $f + g$ and af are defined, we need only show that $\mathcal{L}^2(A)$ is closed under addition and multiplication by a real number.

For addition, we have $(f + g)^2 = f^2 + 2fg + g^2$, so that

$$\int_A (f + g)^2 dm = \int_A f^2 dm + 2\int_A fg dm + \int_A g^2 dm.$$

Since f and g are assumed to be in $\mathcal{L}^2(A)$, $\int_A f^2 dm$ and $\int_A g^2 dm$ are finite. Thus it suffices to show $fg \in \mathcal{L}^2(A)$. This follows from the fact that we require f and g to be measurable and from the inequality $|fg| \leqslant \frac{1}{2}(f^2 + g^2)$, so that $\int_A |fg| dm < \infty$. (See Exercise 26.26.)

For multiplication by real numbers, let $a \in \mathcal{R}$, $f \in \mathcal{L}^2(A)$. Then

$$\int_A (af)^2 dm = a^2 \int_A f^2 dm < \infty. \qquad \square$$

33.3 Corollary: If f and $g \in \mathcal{L}^2(A)$, then $fg \in \mathcal{L}^2(A)$.

33.4 Corollary: If $f \in \mathcal{L}^2(A)$, then $f \in \mathcal{L}(A)$.

Proof: Exercise 34.26. $\qquad \square$

To prove that $\| \ \|_2$ is a norm, we first note that it arises from an inner product: $f \cdot g = \int_A fg dm$; that is, $\|f\|_2 = \sqrt{f \cdot f}$. This is defined and is finite for $f, g \in \mathcal{L}^2(A)$ by Corollary 33.3.

33.5 Proposition: $f \cdot g = \int_A fg dm$ is an inner product on $\mathcal{L}^2(A)$. $\qquad \square$

Proof: Exercise 34.27. Remember that $f = g$ means $f = g$ a.e. on A.

33.6 Corollary: $\| \ \|_2$ is a norm on $\mathcal{L}^2(A)$.

For $\mathcal{L}^2(A)$, the Cauchy inequality and the triangle inequality (also known as the Minkowsky inequality) have the following form.

33.7 Theorem (Cauchy Inequality): For $f, g \in \mathcal{L}^2(A)$, $|f \cdot g| \leqslant \|f\|_2 \|g\|_2$, or

$$\left| \int_A fg dm \right| \leqslant \sqrt{\int_A f^2 dm} \sqrt{\int_A g^2 dm}.$$

33.8 Corollary: For $f, g \in \mathcal{L}^2(A)$, $\int_A |fg| \, dm \leqslant \|f\|_2 \|g\|_2$.

Proof: Exercise 34.28. $\qquad\qquad\qquad\qquad\qquad\qquad\qquad\qquad$ □

33.9 Theorem (Minkowsky inequality): For $f, g \in \mathcal{L}^2(A)$,

$$\|f + g\|_2 \leqslant \|f\|_2 + \|g\|_2,$$

or

$$\sqrt{\int_A (f + g)^2 \, dm} \leqslant \sqrt{\int_A f^2 \, dm} + \sqrt{\int_A g^2 \, dm}.$$

We have shown that $\mathcal{L}^2(A)$ is a real inner product space with $f \cdot g = \int_A fg \, dm$, and norm $\|f\|_2 = \sqrt{\int_A f^2 \, dm}$. We can therefore talk about orthogonal and orthonormal sets (see Example 32.17(3)), and about convergence in the norm. We will delay a discussion of orthogonality until later; it leads in a natural way into the study of Fourier series. But let us explore $\mathcal{L}^2$ convergence now.

Let $\{f_n\}$ be a sequence of functions in $\mathcal{L}^2(A)$, and let $f \in \mathcal{L}^2(A)$. Then, translating our general discussion of convergence in the last section into this context, we say that $\{f_n\}$ converges to f in $\mathcal{L}^2$ (or "in the norm $\| \ \|_2$," or "in the mean") if and only if

$$\lim_{n \to \infty} \|f - f_n\|_2 = 0,$$

i.e.,

$$\lim_{n \to \infty} \int_A (f - f_n)^2 \, dm = 0.$$

What is the relation between $\mathcal{L}^2$ convergence and other notions of convergence we know of for sequences of functions? Uniform convergence surely implies $\mathcal{L}^2$ convergence (Exercise 34.29). Pointwise convergence does not imply $\mathcal{L}^2$ convergence (Exercise 34.30), and $\mathcal{L}^2$ convergence does not imply pointwise convergence (hence does not imply uniform convergence). (See Example 27.2, where $\{f_n\}$ converges in $\mathcal{L}^2$ to 0, but does not converge pointwise for any x. This example shows that $\mathcal{L}^2$ convergence does not even imply pointwise convergence almost everywhere.) On the other hand, if $\{f_n\}$ converges to f in $\mathcal{L}^2$, then a subsequence $\{f_{n_k}\}$ converges to f pointwise a.e.. (See the proof of Theorem 33.11.)

Although $\mathcal{L}^2$ convergence does not imply uniform convergence, it has one important property usually associated with uniform convergence in the theory of the Riemann integral; it is possible to integrate an $\mathcal{L}^2$-convergent sequence term by term.

33.10 Theorem: Given $f_n, f \in \mathcal{L}^2(A)$, if $\lim_{n \to \infty} f_n = f$ in $\mathcal{L}^2$, then

$$\lim_{n \to \infty} \int_A f_n \, dm = \int_A f \, dm.$$

Proof: By Cauchy's Inequality (Theorem 33.7),

$$\left| \int_A f_n \, dm - \int_A f \, dm \right| = \left| \int_A (f_n - f) \, dm \right|$$

$$= \left| \int_A (f_n - f) 1 \, dm \right| \leqslant \| f_n - f \|_2 \| 1 \|_2.$$

Thus, if $\| f_n - f \|_2 \to 0$, then $| \int_A f_n \, dm - \int_A f \, dm | \to 0$. $\qquad\square$

The next question is one of completeness: is $\mathcal{L}^2(A)$ a Hilbert space? The answer is yes. The following theorem is often called the Riesz-Fischer Theorem.

33.11 Theorem: If A is bounded and measurable, then $\mathcal{L}^2(A)$ is complete.*

Proof: Let $\{f_n\}$ be a Cauchy sequence in $\mathcal{L}^2(A)$. We will find a subsequence $\{f_{n_k}\}$ of $\{f_n\}$ which converges pointwise a.e. to a function f. Then we will show that $f \in \mathcal{L}^2(A)$ and that $\{f_n\}$ converges to f in $\mathcal{L}^2$.

For the subsequence, note that since $\{f_n\}$ is Cauchy, given $\epsilon = \frac{1}{2}$, there is an integer n_1 such that for $n > n_1$, $\| f_{n_1} - f_n \|_2 < \frac{1}{2}$. Then there is an $n_2 > n_1$ such that for $n > n_2$, $\| f_{n_2} - f_n \|_2 < \left(\frac{1}{2} \right)^2$. In this way, we obtain a subsequence $\{f_{n_k}\}, k = 1, 2, \ldots$, such that for all k,

$$\| f_{n_k} - f_{n_{k+1}} \|_2 < \left(\frac{1}{2} \right)^k.$$

Letting g be the constant function 1 in the Cauchy Inequality (or rather, in Corollary 33.8), we obtain

$$\int_A | f_{n_k} - f_{n_{k+1}} | \, dm \leqslant \| f_{n_k} - f_{n_{k+1}} \|_2 \| 1 \|_2 < \left(\frac{1}{2} \right)^k \sqrt{m(A)}.$$

---

*The theorem works even if we allow infinite-valued functions (see note for Theorem 28.2). If $f \in \mathcal{L}(A)$, then f is finite a.e. on A.

Thus, $\sum_{k=1}^{\infty} \int_A |f_{n_k} - f_{n_{k+1}}| dm < \sqrt{m(A)}$. By Corollary 28.4 to the Monotone Convergence Theorem, we may interchange the summation and integration, to obtain

$$\int_A \left(\sum_{k=1}^{\infty} |f_{n_k} - f_{n_{k+1}}|\right) dm < \sqrt{m(A)}.$$

Hence $\sum_{k=1}^{\infty} |f_{n_k} - f_{n_{k+1}}| \in \mathcal{L}(A)$, hence is finite a.e. Since absolutely convergent series are convergent, $f(x) = f_{n_1}(x) + \sum_{k=1}^{\infty} (f_{n_{k+1}}(x) - f_{n_k}(x))$ is well defined for almost all $x \in A$. We can let $f(x) = 0$, say, on the exceptional set of measure 0 where the series for $f(x)$ diverges. Note that for almost all $x \in A$,

$$f(x) = f_{n_1}(x) + \lim_{m \to \infty} \sum_{k=1}^{m} (f_{n_{k+1}}(x) - f_{n_k}(x))$$

$$= f_{n_1}(x) + \lim_{m \to \infty} (-f_{n_1}(x) + f_{n_m}(x)) = \lim_{k \to \infty} f_{n_k}(x).$$

Now let us show that $f \in \mathcal{L}^2(A)$. Since $f = (f - f_{n_k}) + f_{n_k}$, and $f_{n_k} \in \mathcal{L}^2(A)$, it suffices to show that $f - f_{n_k} \in \mathcal{L}^2(A)$. Now for $i > k$, $\|f_{n_i} - f_{n_k}\|_2 < \left(\frac{1}{2}\right)^k$, so that $\varlimsup_{i \to \infty} \|f_{n_i} - f_{n_k}\|_2 \leq \left(\frac{1}{2}\right)^k$. Hence by Fatou's Lemma (28.7),

$$\int_A (f - f_{n_k})^2 dm \leq \varliminf_{i \to \infty} \int_A (f_{n_i} - f_{n_k})^2 dm \leq \left(\frac{1}{2}\right)^{2k}.$$

So $f - f_{n_k} \in \mathcal{L}^2(A)$.

Finally, $\{f_n\}$ converges to f in $\mathcal{L}^2$ since

$$\|f - f_n\|_2 \leq \|f - f_{n_k}\|_2 + \|f_{n_k} - f_n\|_2.$$

The first term on the right side is $< \left(\frac{1}{2}\right)^{2k}$, and the second term can be made arbitrarily small (for large enough n, k) since $\{f_n\}$ is a Cauchy sequence. $\qquad \square$

Let us now restrict ourselves to a closed interval $[a,b]$—certainly a bounded measurable set. We know that any continuous function is in $\mathcal{L}^2([a,b])$ (why?). Also, if $f: [a,b] \to \mathcal{R}$ is the $\mathcal{L}^2$ limit of a sequence of continuous functions, then $f \in \mathcal{L}^2([a,b])$ (Exercise 34.33). As we shall prove, the converse is also true; if $f \in \mathcal{L}^2([a,b])$, then f is the $\mathcal{L}^2$ limit of a sequence of continuous functions. Put another way, we say that $C([a,b])$ is *dense* in $\mathcal{L}^2([a,b])$; that is, given $f \in \mathcal{L}^2([a,b])$, and $\epsilon > 0$, there is a function $g \in C([a,b])$ such that $\|f - g\|_2 < \epsilon$. To prove this, we begin, as we often do, with open sets.

33.12 Lemma: If $G \subset [a,b]$ is open and $\epsilon > 0$, then there is a continuous function g such that $\| \chi_G - g \|_2 < \epsilon$; that is, $\int_{[a,b]} (\chi_G - g)^2 dm < \epsilon^2$.

Proof: Write $G = \bigcup_i (a_i, b_i)$ (Theorem 7.2), with the (a_i, b_i) disjoint. We define $g(x) = 0$ for $x \notin G$. For each $i = 1, 2, \ldots$, we attempt to make $\int_{(a_i, b_i)} (\chi_G - g)^2 dm < \epsilon^2/2^i$. Then countable additivity of the integral (Theorem 24.3) will give the result.

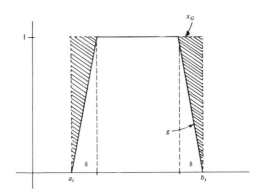

Given (a_i, b_i), therefore, let δ be a positive number less than $\frac{1}{2}(b_i - a_i)$, otherwise undetermined as yet. Then define $g(x) = 1$ for $a_i + \delta \leqslant x \leqslant b_i - \delta$, and let g be linear between the points $(a_i, 0)$ and $(a_i + \delta, 1)$, and between $(b_i - \delta, 1)$ and $(b_i, 0)$. Then g is continuous, and since $0 \leqslant \chi_G - g < 1$ on (a_i, b_i), we have

$$\int_{(a_i, b_i)} (\chi_G - g)^2 dm \leqslant \int_{(a_i, b_i)} (\chi_G - g) dm = 2\left(\frac{1}{2}\delta\right) = \delta.$$

(See shaded triangles in the figure.) Therefore, if we let $\delta \leqslant \epsilon^2/2^i$ in the above construction, our result will follow. ◻

Now we extend our result to measurable sets.

33.13 Lemma: Given $A \subset [a,b]$ measurable, and given $\epsilon > 0$, there is a continuous function g such that $\| \chi_A - g \|_2 < \epsilon$.

Proof: Find an open set $G \subset [a,b]$ such that $A \subset G$ and $m(G) < m(A) + \delta$, where δ is a positive number to be determined. Then, by the previous lemma, there is a continuous function g such that $\| \chi_G - g \|_2 < \epsilon/2$. Therefore,

$$\| \chi_A - g \|_2 \leqslant \| \chi_A - \chi_G \|_2 + \| \chi_G - g \|_2 < \left(\int_{[a,b]} (\chi_G - \chi_A)^2 \, dm \right)^{\frac{1}{2}} + \epsilon/2$$

$$= \left(\int_{[a,b]} (\chi_G - \chi_A) \, dm \right)^{\frac{1}{2}} + \epsilon/2$$

$$= (m(G) - m(A))^{\frac{1}{2}} + \epsilon/2 < \delta^{\frac{1}{2}} + \epsilon/2$$

(Why is $(\chi_G - \chi_A)^2 = (\chi_G - \chi_A)$?) Taking $\delta = (\epsilon/2)^2$ gives the result.
□

33.14 Theorem: Given $f \in \mathcal{L}^2([a,b])$, and given $\epsilon > 0$, there is a continuous function $g: [a,b] \to \mathcal{R}$ such that $\| f - g \|_2 < \epsilon$.

Proof: Using the decomposition $f = f_+ + f_-$, we need only prove the result for f_+ and f_- separately; that is, we may assume $f \geqslant 0$. But in this case, there is a monotone increasing sequence of non-negative simple functions $\{ f_n \}$ which converges pointwise to f (Theorem 19.4). Therefore $g_n = (f - f_n)^2$ defines a sequence, converging pointwise to 0, such that $0 \leqslant g_n \leqslant f^2$. By the Dominated Convergence Theorem (Theorem 28.9), $\lim\limits_{n \to \infty} \int_{[a,b]} g_n \, dm = 0$. Thus $\| f - f_n \|_2 = \sqrt{\int_{[a,b]} g_n \, dm}$ can be made arbitrarily small, for large enough n. Since we have approximated f arbitrarily closely by simple functions, we need only prove the result for simple functions, which are of the form $\sum\limits_{i=1}^{n} c_i \chi_{A_i}$, the A_i being measurable. Therefore, we need only find a continuous function g_i, for each $i = 1, \ldots, n$, such that $\| c_i \chi_{A_i} - g_i \|_2 < \epsilon/n$. This follows from Lemma 33.13. See Exercise 34.35 for details.
□

34. Exercises

34.1 Prove the following for any vector space V.
  (a) $0\overline{u} = \overline{0}$ for all $\overline{u} \in V$.
  (b) $\overline{0}$ is unique; in fact, if $\overline{0}' \in V$ had the property that $\overline{0}' + \overline{u} = \overline{u}$ for some $\overline{u} \in V$, then $\overline{0} = \overline{0}'$.
  (c) Additive inverses are unique; in fact, if $\overline{u} + \overline{v} = \overline{0}$, then $\overline{v} = -\overline{u}$.
  (d) $a\overline{0} = \overline{0}$ for all $a \in \mathcal{R}$.
  (e) $(-1)\overline{u} = -\overline{u}$ for all $\overline{u} \in V$.

34.2 Verify that the Examples 32.2 are all vector spaces.

34.3 Let $M = \{ f: [0,1] \to \mathcal{R} \mid f$ has a maximum and a minimum value$\}$. Prove that M is not a vector space under the ordinary addition and multiplication by a real number.

34.4 Verify that Examples 32.4 are all norms.

34.5 Prove that if $\|\ \|$ is a norm, then

$$\left| \|\bar{u}\| - \|\bar{v}\| \right| \leqslant \|\bar{u} - \bar{v}\|.$$

34.6 Verify that properties (1) – (4) of the metric d follow from the properties of $\|\ \|$.

34.7 Prove the statements in Example 32.6(1) and (2).

34.8 Prove that if $f_n \in C([0,1])$ and $f_n \to f$ in the lub norm (Example 32.4(2)), then $f_n \to f$ in the norm of Example 32.4(3) (called the $\mathcal{L}^1$ norm).

34.9 Prove that any convergent sequence in a normed linear space is a Cauchy sequence.

34.10 Prove that $\mathcal{R}^n$ is complete in its usual norm (Example 32.4(1)). (You may use the fact that $\mathcal{R}$ is complete.)

34.11 Prove that $\mathcal{B}(A)$ is complete in the lub norm (Example 32.4(2)). (Given a Cauchy sequence in this norm, its limit must be the pointwise limit of the sequence.)

34.12 Prove that $C([0,1])$ is *not* complete with the $\mathcal{L}^1$ norm (Example 32.4(3)). (Hint: let $f_n(x) = x^n$.)

34.13 Prove that $\mathcal{R}([0,1])$ is not complete with the $\mathcal{L}^1$ norm (Example 32.4(3)). (Hint: what could $\{f_n\}$ converge to pointwise which would not be in $\mathcal{R}([0,1])$?)

34.14 Prove that if $\bar{u} \cdot \bar{v}$ is an inner product, then $\bar{u} \cdot (\bar{v} + \bar{w}) = \bar{u} \cdot \bar{v} + \bar{u} \cdot \bar{w}$, $\bar{u} \cdot (a\bar{v}) = a(\bar{u} \cdot \bar{v})$, and $\bar{0} \cdot \bar{v} = 0$ for all $\bar{u}, \bar{v}, \bar{w} \in V, a \in \mathcal{R}$.

34.15 Verify that Examples 32.11 are inner products.

34.16 Show that the $\mathcal{L}^1$ norm (Example 32.4(3)) does not come from an inner product by letting

$$f_1(x) = \begin{cases} 0 \text{ on } [0,\frac{1}{2}] \\ 1 \text{ on } (\frac{1}{2},1] \end{cases} \quad \text{and} \quad f_2(x) = \begin{cases} 1 \text{ on } [0,\frac{1}{2}] \\ 0 \text{ on } (\frac{1}{2},1] \end{cases}.$$

Consider $\|f_1 - f_2\|^2$ and $\|f_1 + f_2\|^2$.

34.17 Prove that if $\bar{u} \cdot \bar{v}$ is an inner product, then $\|\bar{u}\| = \sqrt{\bar{u} \cdot \bar{u}}$ has properties (1), (2) and (3) of the norm (Definition 32.3).

34.18 If $\|\bar{u}\| = \sqrt{\bar{u} \cdot \bar{u}}$, show that the Parallelogram Law holds:

$$\|\bar{u} + \bar{v}\|^2 + \|\bar{u} - \bar{v}\|^2 = 2(\|\bar{u}\|^2 + \|\bar{v}\|^2).$$

34.19 (a) Given a quadratic polynomial $p(x) = ax^2 + bx + c$, if $p(x) \geqslant 0$ for all x, show that the discriminant of p (namely $b^2 - 4ac$) is $\leqslant 0$.

(b) Give a proof of the Cauchy Inequality by considering

$$p(x) = (\bar{u} + x\bar{v}) \cdot (\bar{u} + x\bar{v}),$$

and using part (a).

34.20 Verify the statements of Example 32.17. (Hint for (3):

$$\sin(a + b) + \sin(a - b) = 2\sin a \cos b$$

$$\cos(a + b) + \cos(a - b) = 2\cos a \cos b$$

$$\cos(a - b) - \cos(a + b) = 2\sin a \sin b.)$$

34.21 (a) Prove that an orthogonal set in an inner product space is linearly independent.

(b) Prove that if $\{\bar{v}_1, \ldots, \bar{v}_n\}$ is an orthonormal basis for V, then any $\bar{v} \in V$ can be expressed as $\bar{v} = \sum_{i=1}^{n} (\bar{v} \cdot \bar{v}_i)\bar{v}_i$.

34.22 Let

$$f_0(x) = \begin{cases} 1 \text{ if } 2n + 1 < x < 2n \text{ for some integer } n \\ -1 \text{ if } 2n < x < 2n + 1 \text{ for some integer } n \\ 0 \text{ otherwise.} \end{cases}$$

Let $f_n(x) = f_0(2^n x)$ for $n = 1, 2, \ldots$. Prove that the set $\{f_0, f_1, \ldots\}$ is orthonormal in $\mathcal{L}([0,1])$ with respect to the inner product

$$f \cdot g = \int_{[0,1]} fg \, dm.$$

34.23 (a) Given a point (x_0, y_0) and a line $ax + by = c$ in $\mathfrak{R}^2$, use the ideas of section 32 to find the point on the line closest to (x_0, y_0), and the distance from (x_0, y_0) to the line.

(b) Repeat (a) for a point and a plane in $\mathfrak{R}^3$.

34.24 Find a non-measurable function f such that f^2 is measurable.

34.25 Show that the relation "equal a.e." is an equivalence relation.

34.26 Prove Corollary 33.4.

34.27 Prove Proposition 33.5.

34.28 Prove Corollary 33.8. (Hint: let

$$h(x) = \begin{cases} 1 \text{ if } f(x)g(x) \geqslant 0 \\ -1 \text{ if } f(x)g(x) < 0. \end{cases}$$

Then $fgh = |fg|$. Apply Theorem 33.7.)

34.29 Prove that if $f_n, f \in \mathcal{L}^2(A)$ and $f_n \to f$ uniformly on A, then $f_n \to f$ in $\mathcal{L}^2$.

34.30 Show that pointwise convergence does not imply $\mathcal{L}^2$ convergence. (See section 27.)

34.31 (a) Suppose that f is bounded and measurable on a bounded measurable set A. Prove that $f \in \mathcal{L}^2(A)$.
(b) Let $f(x) = 1/\sqrt{|x|}$ for $x \neq 0$, $f(0) = 0$. Show

$$f \in \mathcal{L}([-1,1]) \setminus \mathcal{L}^2([-1,1]).$$

34.32 Let $f_n, f, g \in \mathcal{L}^2(A)$, and $f_n \to f$ in $\mathcal{L}^2$. Prove that $\lim_{n \to \infty} \int_A f_n g \, dm = \int_A fg \, dm$.

34.33 If $f_n \in \mathcal{L}^2(A)$ for all n, and $f_n \to f$ in $\mathcal{L}^2$, then show that $f \in \mathcal{L}^2(A)$.

34.34 If $f_n \in \mathcal{L}^2(A)$, $f_n \to f$ in $\mathcal{L}^2$, and g is bounded and measurable, then $f_n g \to fg$ in $\mathcal{L}^2$.

34.35 Give all of the details of the proof of Theorem 33.14.

34.36 Show that the unit sphere of $\mathcal{L}^2([0,1])$ is not compact. (The unit sphere is $\{f \in \mathcal{L}^2([0,1]) \mid \|f\| = 1\}$. Find a sequence g_n such that $\|g_n\| = 1$ and $\|g_n - g_k\| \geqslant 1$ for $n \neq k$.)

34.37 For $f \in \mathcal{L}(A)$, define the $\mathcal{L}^1$ norm by $\|f\|_1 = \int_A |f| \, dm$. As for $\mathcal{L}^2(A)$, we identify functions which are equal a.e. on A.
(a) Show that $\| \ \|_1$ is a norm.
(b) Prove that $\mathcal{L}(A)$ is complete in the norm $\| \ \|_1$.
(c) Show that $\mathcal{C}(A)$ is dense in $\mathcal{L}(A)$ in norm $\| \ \|_1$.

34.38 Generalize 34.37(c) to show that if $\mathcal{D} \subset \mathcal{L}^2(A)$ is dense in $\mathcal{L}^2(A)$ in norm $\| \ \|_2$, then $\mathcal{D}$ is dense in $\mathcal{L}(A)$ in norm $\| \ \|_1$. (Hint: show that $\mathcal{L}^2(A)$ is dense in $\mathcal{L}(A)$ in norm $\| \ \|$.)

34.39 Suppose that $fg \in \mathcal{L}([a,b])$ for all $f \in \mathcal{L}^2([a,b])$. Show that $g \in \mathcal{L}^2([a,b])$.

34.40 Let $\sum_{n=1}^{\infty} \|f_n\|_2 < \infty$. Show that $\sum_{n=1}^{\infty} f_n$ converges absolutely a.e.,

$$f = \sum_{n=1}^{\infty} f_n \in \mathcal{L}^2, \quad \text{and} \quad \|f\|_2 \leqslant \sum_{n=1}^{\infty} \|f_n\|_2.$$

34.41 (a) Show that $\|f\|_1 \leqslant \|f\|_2\sqrt{m(A)}$ for $f \in \mathcal{L}^2(A)$. Hence if $f_n \to f$ in $\mathcal{L}^2$, then $f_n \to f$ in $\mathcal{L}^1$.

    (b) Find a sequence $\{f_n\}$ which converges to 0 in $\mathcal{L}^1$ but does not converge in $\mathcal{L}^2$.

34.42 If A is bounded and measurable and p a positive real number, define $\mathcal{L}^p(A) = \{f:A \to \Re \mid f$ is measurable and $\int_A |f|^p dm < \infty\}$. Note that $\mathcal{L}^2$ and $\mathcal{L}^1$ are special cases.

    (a) For $f \in \mathcal{L}^p(A)$, define $\|f\|_p = \sqrt[p]{\int_A |f|^p}$. As in $\mathcal{L}^2$ we consider two functions equivalent if they are equal almost everywhere. Show that properties (1) – (3) of definition 32.3 hold for $\mathcal{L}^p(A)$. The remainder of this exercise will deal with proving property (4) of 32.3 for $\mathcal{L}^p(A)$ when $p \geqslant 1$.

    (b) Prove that $a^\lambda b^{1-\lambda} \leqslant \lambda a + (1 - \lambda)b$ for $0 < \lambda < 1$ and a and b non-negative real numbers. (Hint: Take the log of both sides of the inequality and use the concavity of the graph of log).

    (c) Prove Hölder's Inequality: if p and q are real numbers greater than 1 such that $1/p + 1/q = 1$ and if $f \in \mathcal{L}^p(A)$ and $g \in \mathcal{L}^q(A)$, then $fg \in \mathcal{L}^1(A)$ and $\int_A |fg| < \|f\|_p \|g\|_q$. (Hint: let $|f|^p = a$ and $|g|^q = b$ and $\lambda = 1/p$ in part (b) above).

    (d) Prove property (4) of 32.3 for $\mathcal{L}^p(A)$ with $p > 1$. This is known as Minkowski's Inequality and states that $\|f + g\|_p < \|f\|_p + \|g\|_p$. (Hint: note that $(f + g)^p = f(f + g)^{p-1} + g(f + g)^{p-1}$, and apply Hölder's Inequality to each term of the right hand side.)

    (e) Show that $\mathcal{L}^p(A)$ is a real linear space.

    It turns out that these $\mathcal{L}^p$ spaces are complete, but it is beyond our present scope to show this.

CHAPTER 8

The L^2 Theory of Fourier Series

35. Definition and Examples

The study of Fourier Series amounts to approximating certain functions by *trigonometric polynomials* of the form $\sum_{k=0}^{n} (a_k \cos kx + b_k \sin kx)$ or *trigonometric series* of the form

$$\sum_{k=0}^{\infty} (a_k \cos kx + b_k \sin kx).$$

Because the sine and cosine functions are useful in many physical applications involving waves (e.g. spectroscopy, propagation of sound waves, heat equations), the study of Fourier Analysis is of great importance to physical scientists. In the next Chapter we will discuss one of these applications. For now we restrict ourselves to $\mathcal{L}^2$ where the study of Fourier Series leads to particularly nice results. We shall also restrict our attention to the interval $[-\pi, \pi]$ since trigonometric functions are periodic of period 2π (i.e., $\sin x = \sin(x + 2\pi)$ for all x).

Recall from Chapter 7 that $\mathcal{L}^2[-\pi, \pi]$ has an inner product defined by

$$f \cdot g = \int_{[-\pi,\pi]} fg \, dm$$

and the resulting norm

$$\|f\|_2 = \sqrt{\int_{[-\pi,\pi]} f^2 dm}.$$

Also recall that $\{u_1, \cdots, u_n, \cdots\}$ is an orthonormal set in $\mathcal{L}^2$ if and only if $u_i \cdot u_j = 0$ whenever $i \neq j$, and $\|u_i\|_2 = 1$ for every i. There are many orthonormal sets in $\mathcal{L}^2[-\pi, \pi]$ but in Fourier Analysis one uses the set

$$\left\{ \frac{1}{\sqrt{2\pi}}, \frac{1}{\sqrt{\pi}} \sin kx, \frac{1}{\sqrt{\pi}} \cos kx \right\}_{k=1}^{\infty}$$

(see example 32.17 and exercise 34.20). Many of our results hold for any (countable) orthonormal set in $\mathcal{L}^2$, and we will often use the more general notation $\{u_i\}_{i=1}^{\infty}$.

35.1 Lemma:

$$\left\{ \frac{1}{\sqrt{2\pi}}, \frac{1}{\sqrt{\pi}} \sin kx, \frac{1}{\sqrt{\pi}} \cos kx \right\}_{k=1}^{\infty}$$

is an orthonormal set in $\mathcal{L}^2[-\pi, \pi]$.

Note: In doing Fourier Analysis, it is often helpful (and traditional) to use the Riemann notation for the Lebesgue integral. That is, we will write $\int_a^b f(x)dx$ instead of $\int_{[a,b]} f dm$, even though f may not be Riemann integrable.

Proof: First,

$$\int_{-\pi}^{\pi} \frac{1}{\sqrt{2\pi}} \frac{\cos kx}{\sqrt{\pi}} dx = \frac{1}{\pi\sqrt{2}} \int_{-\pi}^{\pi} \cos kx\, dx = \frac{\sin kx}{k\pi\sqrt{2}} \bigg|_{-\pi}^{\pi} = 0,$$

and similarly

$$\frac{1}{\sqrt{2\pi}} \int_{-\pi}^{\pi} \sin kx\, dx = 0.$$

Also, using the trigonometric identity

$$\cos A \cos B = \frac{1}{2}\cos(A+B) + \frac{1}{2}\cos(A-B),$$

$$\int_{-\pi}^{\pi} \frac{\cos mx \cos nx}{\sqrt{\pi}\sqrt{\pi}} dx = \frac{1}{2}\int_{-\pi}^{\pi} \frac{\cos(m+n)x}{\pi} dx + \frac{1}{2}\int_{-\pi}^{\pi} \frac{(\cos(m-n)x}{\pi} dx$$

which $= 0$ if $m \neq n$ and $= 1$ if $m = n$ (Verify!)

Similarly

$$\int_{-\pi}^{\pi} \frac{\sin mx}{\sqrt{\pi}} \frac{\sin nx}{\sqrt{\pi}} dx = \begin{cases} 0 \text{ if } m \neq n \\ 1 \text{ if } m = n \end{cases}$$

and

$$\int_{-\pi}^{\pi} \frac{\sin mx}{\sqrt{\pi}} \frac{\cos nx}{\sqrt{\pi}} dx = 0$$

for all m, n. $\square$

If you have any trouble (Riemann) integrating in the above proof, see Exercise 34.20 of Chapter 7.

Given a finite orthonormal set $\{u_1, \dots, u_n\}$, Theorem 32.18 says that the best approximation of f by a sum of the form $\sum_{k=1}^{n} c_k u_k$ is the projection of f on the subspace spanned by $\{u_1, \dots, u_n\}$. Therefore, for a countable orthonormal set $\{u_k\}_{k=1}^{\infty}$, we are motivated to try to approximate f by $\sum_{k=1}^{\infty} c_k u_k$ where $c_k = f \cdot u_k$. That is, we suspect that the best approximation for f in terms of the u_k (in the $\mathcal{L}^2$ norm) is obtained by using coefficients of the form $c_k = f \cdot u_k$. We give these a special name.

35.2 Definition: Let $\{u_k\}_{k=1}^{\infty}$ be an orthonormal set in $\mathcal{L}^2[-\pi, \pi]$ and let $f \in \mathcal{L}^2$. Then the nth *generalized Fourier coefficient of f with respect to the set* $\{u_k\}_{k=1}^{\infty}$ *is* $c_n = f \cdot u_n = \int_{[-\pi,\pi]} f \cdot u_n dm = \int_{-\pi}^{\pi} f(x) \cdot u_n(x) dx.$ *

The *generalized Fourier series of f with respect to the orthonormal set* is

$$\sum_{k=1}^{\infty} c_k u_k.$$

Throughout the remainder of the text we will be primarily interested in the particular orthonormal set

$$\left\{ \frac{1}{\sqrt{2\pi}}, \frac{\sin kx}{\sqrt{\pi}}, \frac{\cos kx}{\sqrt{\pi}} \right\}_{k=1}^{\infty}.$$

---

*Note that this inner product is always defined for $f \in \mathcal{L}^2$ (see 33.3).

In this case we drop the word "generalized," and the Fourier coefficients are

$$\alpha_0 = \int_{-\pi}^{\pi} f(x) \frac{1}{\sqrt{2\pi}} dx, \quad \alpha_k = \int_{-\pi}^{\pi} f(x) \frac{\cos kx}{\sqrt{\pi}} dx,$$

$$\beta_k = \int_{-\pi}^{\pi} f(x) \frac{\sin kx}{\sqrt{\pi}} dx, \quad k > 0.$$

The *Fourier Series* associated with f is

$$\alpha_0 \frac{1}{\sqrt{2\pi}} + \sum_{k=1}^{\infty} \left(\alpha_k \frac{\cos kx}{\sqrt{\pi}} + \beta_k \frac{\sin kx}{\sqrt{\pi}} \right).$$

It is conventional to write the Fourier Series for f in the form

$$\frac{a_0}{2} + \sum_{k=1}^{\infty} (a_k \cos kx + b_k \sin kx)$$

where

$$a_0 = \frac{1}{\pi} \int_{-\pi}^{\pi} f(x) dx, \quad a_k = \frac{1}{\pi} \int_{-\pi}^{\pi} f(x) \cos kx \, dx, \quad b_k = \frac{1}{\pi} \int_{-\pi}^{\pi} f(x) \sin kx \, dx$$

and we will adopt this practice. It is easy to see that the same series results. Note however that $\{\frac{1}{2}, \cos kx, \sin kx\}_{k=1}^{\infty}$ is not an orthonormal system so that this alternative method is merely a notational convenience.

The most interesting questions about Fourier Series have to do with convergence. Given a function $f \in \mathcal{L}^2[-\pi, \pi]$, we have shown how to obtain its Fourier Series, but we do not know whether this series converges (uniformly, $\mathcal{L}^2$, or at a point) and if it converges at x whether it converges to the value $f(x)$. The remainder of the text will deal with these questions. For now let us look at some examples of Fourier Series.

35.3 Example: Let

$$f(x) = \begin{cases} 0 & \text{for } -\pi \leqslant x < 0 \\ 1 & \text{for } 0 \leqslant x \leqslant \pi. \end{cases}$$

Then

$$a_k = \frac{1}{\pi} \int_{-\pi}^{\pi} f(x) \cos kx \, dx = \frac{1}{\pi} \int_{0}^{\pi} \cos kx \, dx = \begin{cases} \left. \frac{1 \sin kx}{\pi k} \right|_0^{\pi} = 0 \text{ if } k \neq 0 \\ \left. \frac{1}{\pi} x \right|_0^{\pi} = 1 \text{ if } k = 0. \end{cases}$$

Similarly,

$$b_k = -\frac{1}{\pi}\frac{\cos kx}{k}\Big|_0^\pi = \begin{cases} \dfrac{1}{\pi}\cdot\dfrac{2}{k} \text{ for } k \text{ odd}\\[2mm] 0 \text{ for } k \text{ even}.\end{cases}$$

Thus the Fourier Series associated with f is

$$\frac{1}{2}+\frac{2}{\pi}\sum_{k=1}^\infty \frac{\sin(2k-1)x}{2k-1} = \frac{1}{2}+\frac{2}{\pi}\left[\sin x + \frac{\sin 3x}{3} + \frac{\sin 5x}{5} + \cdots\right].$$

Notice that $f(0) = 1$, but at 0 the Fourier Series equals $\frac{1}{2}$. In fact, since we are using Lebesgue integration, any functions f and g equivalent in $\mathcal{L}^2$ (i.e. $f(x) = g(x)$ for a.e. x) will have identical Fourier Series. This brief discussion should prepare the reader for some of the problems involving pointwise convergence of Fourier Series which we present in Chapter 9.

35.4 Example: Let $f = \chi_Q$ on $[-\pi,\pi]$, the characteristic function of the rationals. Then

$$\frac{1}{\pi}\int_{-\pi}^\pi f(x)\sin kx\,dx = 0 \quad \text{and} \quad \frac{1}{\pi}\int_{-\pi}^\pi f(x)\cos kx\,dx = 0$$

for all k (why?). Thus the Fourier Series for f is identically 0. Notice that $f(x)$ does not equal its Fourier Series at each rational number x.

35.5 Example: Let $f(x) = |x|$ on $[-\pi,\pi]$. Then each b_k is 0 because f is an even function (see Exercise 38.1). Also $a_0 = \pi$ (verify) and

$$a_k = \frac{1}{\pi}\int_{-\pi}^\pi |x|\cos kx\,dx = \frac{2}{\pi}\int_0^\pi x\cos kx\,dx = \begin{cases} -\dfrac{4}{\pi k^2} \text{ for } k \text{ odd}\\[2mm] 0 \quad\text{ for } k \text{ even}.\end{cases}$$

Thus the Fourier Series for f is

$$\frac{\pi}{2} - \frac{4}{\pi}\left[\cos x + \frac{\cos 3x}{9} + \frac{\cos 5x}{25} + \cdots\right].$$

More examples are to be found in the exercises at the end of this Chapter.

36. Elementary Properties

We first restate a result from Chapter 7 which shows that the generalized Fourier coefficients of f yield the best approximation of f in $\mathcal{L}^2[-\pi, \pi]$. Although we are primarily interested in the orthonormal set

$$\left\{\frac{1}{\sqrt{2\pi}}, \frac{\sin kx}{\sqrt{\pi}}, \frac{\cos kx}{\sqrt{\pi}}\right\}_{k=1}^{\infty},$$

we will use the general notation for clarity.

36.1 Theorem: Let $\{u_k\}_{k=1}^{\infty}$ be an orthonormal set in $\mathcal{L}^2[-\pi, \pi]$. Define, for $f \in \mathcal{L}^2[-\pi, \pi]$, $c_k = \int_{[-\pi,\pi]} f u_k dm$ and $s_n(x) = \sum_{k=0}^{n} c_k u_k(x)$ [s_n is called the nth generalized Fourier approximation of f.]. Let $t_n(x) = \sum_{k=0}^{n} d_k u_k(x)$ where the coefficients are arbitrary.

Then $s_n(x)$ is a better $\mathcal{L}^2$ approximation of f than $t_n(x)$. That is,

$$\int_{[-\pi,\pi]} (f - s_n)^2 dm \leqslant \int_{[-\pi,\pi]} (f - t_n)^2 dm.$$

Proof: This is just Theorem 32.18 of Chapter 7. □

36.2 Corollary: With the above hypotheses $\int_{[-\pi,\pi]} f^2 dm \geqslant \sum_{k=0}^{n} c_k^2$ for every n.

Proof: See Corollary 32.19. □

36.3 Corollary: (Bessel's Inequality) $\sum_{k=0}^{\infty} c_k^2 \leqslant \int_{[-\pi,\pi]} f^2 dm$.

Proof: This follows since Corollary 36.2 above holds for all n. □

36.4 Corollary: (Riemann-Lebesgue Lemma) $\lim_{k \to \infty} c_k = 0$.

Proof: $\int_{[-\pi,\pi]} f^2 dm$ bounds the positive term series $\sum_{k=0}^{\infty} c_k^2$ by Corollary 36.3, so the series converges and the result follows. □

36.5 Corollary: In the $\mathcal{L}^2$ sense, successive generalized Fourier approximations improve as n increases (i.e. $\int_{[-\pi,\pi]} (f - s_n)^2 dm \geqslant \int_{[-\pi,\pi]} (f - s_{n+1})^2 dm$).

Proof: This follows from Theorem 36.1 since

$$\int_{[-\pi,\pi]} (f - s_{n+1})^2 dm \leqslant \int_{[-\pi,\pi]} (f - t_n)^2 dm$$

where $t_n = s_n + 0u_{n+1}$. □

36.6 Corollary: For the orthonormal set

$$\left\{ \frac{1}{\sqrt{2\pi}}, \frac{\sin kx}{\sqrt{\pi}}, \frac{\cos kx}{\sqrt{\pi}} \right\}_{k=1}^{\infty}$$

(1) Bessel's Inequality becomes

$$\frac{a_0}{2} + \sum_{k=1}^{\infty} (a_k^2 + b_k^2) \leqslant \frac{1}{\pi} \int_{-\pi}^{\pi} f^2 dm$$

where

$$a_k = \frac{1}{\pi} \int_{-\pi}^{\pi} f(x) \cos kx\, dx \quad \text{and} \quad b_k = \frac{1}{\pi} \int_{-\pi}^{\pi} f(x) \sin kx\, dx,$$

and (2) Riemann-Lebesgue becomes

$$\lim_{k \to \infty} \int_{-\pi}^{\pi} f(x) \cos kx\, dx = \lim_{k \to \infty} \int_{-\pi}^{\pi} f(x) \sin kx\, dx = 0.$$

Bessel's Inequality shows that an $\mathcal{L}^2$ function f has Fourier coefficients which obey

$$\frac{a_0}{2} + \sum_{k=1}^{\infty} (a_k^2 + b_k^2) \leqslant \frac{1}{\pi} \|f\|_2^2 < \infty.$$

That is, if $f \in \mathcal{L}^2$, then

$$\frac{a_0^2}{2} + \sum_{k=1}^{\infty} (a_k^2 + b_k^2)$$

converges. The converse is also true: if there are numbers $\{a_i\}_{i=0}^{\infty}$ and $\{b_i\}_{i=1}^{\infty}$ such that

$$\frac{a_0^2}{2} + \sum_{k=1}^{\infty} (a_k^2 + b_k^2)$$

converges, then these numbers are the Fourier coefficients of some function $f \in \mathcal{L}^2$. The key to the proof is the completeness* of $\mathcal{L}^2$ (see Theorem

*For this reason this result is often called the Riesz-Fischer Theorem.

33.11) and the fact that $\mathcal{L}^2$ convergence permits integration term by term (Theorem 33.10). We prove the theorem for any (countable) orthonormal set which of course includes the special case of trigonometric series.

36.6 Theorem: (Riesz-Fischer Theorem) Let $\{u_i\}_{i=1}^{\infty}$ be an orthonormal set in $\mathcal{L}^2[-\pi,\pi]$ and suppose $\sum\limits_{i=1}^{\infty} c_i^2$ is a convergent set of real numbers.

Then there exists a function $f \in \mathcal{L}^2[-\pi,\pi]$ such that $c_i = f \cdot u_i$ and the partial sums $s_n = c_1 u_1 + \cdots + c_n u_n$ converge to f in the $\mathcal{L}^2$ sense.

Proof: For $n > m$, $(\|s_n - s_m\|_2)^2 = c_{m+1}^2 + \cdots + c_n^2$ by orthonormality of the u_i (calculate this!). Thus $\{s_n\}$ is a Cauchy sequence in $\mathcal{L}^2$ so that by completeness of $\mathcal{L}^2$(33.11) there is an $f \in \mathcal{L}^2$ such that

$$\lim_{n \to \infty} \|f - s_n\|_2 = 0.$$

For $n > i$,

$$c_i = s_n \cdot u_i = \int_{[-\pi,\pi]} s_n u_i \, dm,$$

and

$$\lim_{n \to \infty} \int_{[-\pi,\pi]} s_n u_i \, dm = \int_{[-\pi,\pi]} f u_i \, dm$$

by Exercise 34.32 of Chapter 7. So $c_i = \int_{[-\pi,\pi]} f u_i \, dm$. That is, the c_i's are indeed the generalized Fourier coefficients of f. □

37. $\mathcal{L}^2$ Convergence of Fourier Series

We have seen that Fourier Series provide a means of approximating a function $f \in \mathcal{L}^2$. In fact, given an orthonormal set $\{u_i\}_{i=1}^{\infty}$ in $\mathcal{L}^2[-\pi,\pi]$, the generalized Fourier coefficients $c_k = \int_{[-\pi,\pi]} f u_k \, dm$ yield partial sums $s_n = \sum\limits_{k=1}^{n} c_k u_k$ such that $\|f - s_n\|_2 \leqslant \|f - t_n\|_2$ where t_n is any other series of the form $\sum\limits_{k=1}^{n} d_k u_k$. It follows (Corollary 36.5) that successive partial Fourier sums get closer and closer to f in the $\mathcal{L}^2$ sense. We would hope in addition, that

$$\lim_{n \to \infty} s_n = f \text{ in the sense of } \mathcal{L}^2.$$

In fact this is true. It takes a fair amount of work to reach this result, however, all of it interesting in its own right. We begin with a definition.

37.1 Definition: A series of numbers $\sum\limits_{k=1}^{\infty} a_k$ is said to have $(C,1)$ *sum (or Cesàro sum)* s if $\lim\limits_{n \to \infty} \sigma_n = s$ where

$$\sigma_n = \frac{s_1 + s_2 + \cdots + s_n}{n} \quad \text{and} \quad s_n = \sum_{k=1}^{n} a_k.$$

($\{\sigma_n\}$ is called the sequence of arithemetic means.)

37.2 Example: Many divergent series converge $(C,1)$. For example,

$$1 - 1 + 1 - 1 + \cdots$$

diverges, but converges to $\frac{1}{2}$ $(C,1)$ since $\sigma_n = \frac{1}{2}$ when n is even and $\sigma_n = \frac{n+1}{2n}$ when n is odd (see Exercise 38.7).

37.3 Example: If a series Σa_k converges to the real number s, then it also converges $(C,1)$ to s. Observe that

$$|s - \sigma_n| = \frac{1}{n}|(s - s_1) + \cdots + (s - s_n)| \leqslant \frac{1}{n}|s - s_1| + \cdots + \frac{1}{n}|s - s_n|.$$

Given $\epsilon > 0$, there is an N such that for $k > N$, $|s - s_k| < \epsilon$. Also there exists an M such that $|s - s_k| < M$ for $k \leqslant N$. Thus for $n > N$,

$$|s - \sigma_n| < \frac{N}{n}M + \frac{n-N}{n}\epsilon.$$

Thus $|s - \lim\limits_{n \to \infty} \sigma_n| \leqslant \epsilon$.

We will obtain the remarkable and useful result that for a continuous function, the Fourier partial sums converge *uniformly* in the $(C,1)$ sense. (Note that there may be points at which the Fourier Series may not even converge in the usual sense—see the next chapter.)

We now need two unexciting, but necessary trigonometric identities.

37.4 Lemma: (1) For $\theta \neq 2n\pi$ for any integer n,

$$\frac{1}{2} + \sum_{k=1}^{n} \cos k\theta = \frac{\sin\left(n + \frac{1}{2}\right)\theta}{2\sin\left(\frac{\theta}{2}\right)}.$$

(2) For $\theta \neq n\pi$ for any integer n,

$$\sin\theta + \sin 3\theta + \cdots + \sin(2n-1)\theta = \frac{\sin^2 n\theta}{\sin\theta}.$$

Proof: Using the trigonometric identity

$$\sin(B+A) - \sin(B-A) = 2\sin A \cos B,$$

and letting $B = k\theta$ and $A = \frac{\theta}{2}$, we obtain

$$\sin\left(k+\frac{1}{2}\right)\theta - \sin\left(k-\frac{1}{2}\right)\theta = 2\sin\left(\frac{1}{2}\theta\right)\cos k\theta.$$

Letting $k = 0,1,\cdots,n$ and adding, we obtain

$$\sin\left(n+\frac{1}{2}\right)\theta + \sin\left(\frac{\theta}{2}\right) = 2\sin\left(\frac{\theta}{2}\right)[1 + \cos\theta + \cos 2\theta + \cdots + \cos n\theta]$$

from which part (1) follows easily.

(2) Using the relation

$$\sin A \sin B = \frac{\cos(A-B) - \cos(A+B)}{2}$$

we obtain

$$\sin\theta[\sin\theta + \sin 3\theta + \cdots + \sin(2n-1)\theta] = \frac{1 - \cos 2n\theta}{2}.$$

Using the cosine double angle formula, this equals $\sin^2 n\theta$. □

We are now ready to prove some important results about trigonometric series. The ultimate goal is to show that for $f \in \mathcal{L}^2[-\pi,\pi]$, the partial sums s_n converge to f in the $\mathcal{L}^2$ sense. On the way we will prove that for f continuous, the $(C,1)$ sums σ_n of the Fourier Series converge uniformly to f. To achieve these results we need the following integral representations of $s_n(x)$ and $\sigma_n(x)$.

Note: From now on we will find it necessary to use expressions such as $f(x+t)$, where $f \in \mathcal{L}^2[-\pi,\pi]$ and $x \in [-\pi,\pi]$. Since $x+t$ need not be in $[-\pi,\pi]$, f must be defined outside this interval. In this case, we assume f is extended periodically (with period 2π) to all of $\mathcal{R}$. That is, $f(x+2\pi) = f(x)$ for all $x \in \mathcal{R}$. Of course, this may make it necessary to change the value at either $-\pi$ or π, but this will not affect any integrals involving f.

37.5 Lemma: Let $f \in \mathcal{L}^2[-\pi,\pi]$.* Then the Fourier partial sums

$$s_n(x) = \frac{1}{\pi} \int_0^\pi [f(x+t) + f(x-t)] D_n(t) dt,$$

where

$$D_n(t) = \begin{cases} 0 & \text{if } t = 2n\pi, \ n \text{ an integer} \\ \dfrac{\sin\left(n+\dfrac{1}{2}\right)t}{2\sin\left(\dfrac{t}{2}\right)} & \text{otherwise.} \end{cases}$$

(D_n is called the *Dirichlet Kernel* of order n. Note that $D_n(-t) = D_n(t)_n$).

Proof: By definition,

$$s_n(x) = \frac{a_0}{2} + \sum_{k=1}^{n} (a_k \cos kx + b_k \sin kx)$$

$$= \frac{1}{2\pi} \int_{-\pi}^{\pi} f(t) dt + \frac{1}{\pi} \sum_{k=1}^{n} \left(\cos kx \int_{-\pi}^{\pi} f(t) \cos kt \, dt + \sin kx \int_{-\pi}^{\pi} f(t) \sin kt \, dt \right)$$

$$= \frac{1}{\pi} \int_{-\pi}^{\pi} f(t) \left[\frac{1}{2} + \sum_{k=1}^{n} \cos kx \cos kt + \sin kx \sin kt \right] dt$$

$$= \frac{1}{\pi} \int_{-\pi}^{\pi} f(t) \left[\frac{1}{2} + \sum_{k=1}^{n} \cos k(x-t) \right] dt$$

$$\overset{37.4}{=} \frac{1}{\pi} \int_{-\pi}^{\pi} f(t) \left[\frac{\sin\left(n+\dfrac{1}{2}\right)(x-t)}{2\sin\left(\dfrac{x-t}{2}\right)} \right] dt.**$$

$\square$

*Actually, $f \in \mathcal{L}[-\pi,\pi]$ is enough here.
**This integrand is not defined for $x - t = 2m\pi, m$ an integer. However, this occurs only on a set of measure zero, so that the integral exists.

By exercise 26.35 and exercise 38.14,

$$s_n(x) = \frac{1}{\pi}\int_{x-\pi}^{x+\pi} f(x-u)\left[\frac{\sin\left(n+\frac{1}{2}\right)u}{2\sin\left(\frac{u}{2}\right)}\right]du$$

$$= \frac{1}{\pi}\int_{-\pi}^{\pi} f(x-u)D_n(u)du = \frac{1}{\pi}\int_{-\pi}^{0} f(x-u)D_n(u)du + \frac{1}{\pi}\int_{0}^{\pi} f(x-u)D_n(u)du$$

$$= \frac{1}{\pi}\int_{0}^{\pi} f(x+t)D_n(-t)dt + \frac{1}{\pi}\int_{0}^{\pi} f(x-t)D_n(t)dt$$

$$= \frac{1}{\pi}\int_{0}^{\pi} [f(x+t)+f(x-t)]D_n(t)dt. \qquad \square$$

37.6 Corollary: Let

$$\sigma_n(x) = \frac{s_0(x)+\cdots+s_{n-1}(x)}{n}$$

for $s_n(x)$ defined as in the Lemma. Then

$$\sigma_n(x) = \frac{1}{\pi}\int_{0}^{\pi} [f(x+t)+f(x-t)]K_n(t)dt,$$

where

$$K_n(t) = \begin{cases} 0 & \text{for } t = 2m\pi, m \text{ an integer} \\ \dfrac{\sin^2\left(\dfrac{nt}{2}\right)}{2n\sin^2\left(\dfrac{t}{2}\right)} & \text{otherwise.} \end{cases}$$

$(K_n$ is called the *Fejer Kernel* of order n.)

Proof: Clearly,

$$\sigma_n(x) = \frac{1}{n}\sum_{k=0}^{n-1} \frac{1}{\pi}\int_{0}^{\pi} [f(x+t)+f(x-t)]D_k(t)dt,$$

so we need only show that

$$\frac{1}{n}\sum_{k=0}^{n-1} D_k(t) = \frac{\sin^2(nt/2)}{2n\sin^2(t/2)}.$$

But

$$\frac{1}{n}\sum_{k=0}^{n-1}D_k(t) = \frac{1}{2n\sin(t/2)}\sum_{k=0}^{n-1}\sin\left(k+\frac{1}{2}\right)t$$

by definition of $D_k(t)$, and

$$\sum_{k=0}^{n-1}\sin\left(k+\frac{1}{2}\right)t = \sum_{j=1}^{n}\sin\left(j-\frac{1}{2}\right)t = \sum_{j=1}^{n}\sin(2j-1)\frac{t}{2} = \frac{\sin^2(nt/2)}{\sin(t/2)}$$

by 37.4.   □

We are nearly in a position to prove the desired result about $(C,1)$ convergence. All we need are two easy results concerning the Fejer Kernel $K_n(t)$.

37.7 Lemma: (1) $$\frac{2}{\pi}\int_{[0,\pi]}K_n(t)dt = 1$$

for $n = 1,2\cdots$ and

(2) if t satisfies $0 < \delta \leqslant |t| \leqslant \pi$, then

$$K_n(t) \leqslant \frac{1}{2n\sin^2(\delta/2)}.$$

Proof: (1) Apply Corollary 37.6 to $f(x) \equiv 1$, observing that all $s_n(x) = 1$.

(2) Use the definition

$$K_n(t) = \frac{\sin^2(nt/2)}{2n\sin^2(t/2)}$$

and note that on $[\delta,\pi]$ the minimum value of $\sin^2(t/2)$ is obtained when $t = \delta$.   □

37.8 Theorem: Let f be a continuous function on $[-\pi,\pi]$ (thus $f \in \mathcal{L}^2[-\pi,\pi]$). Then σ_n converges to f uniformly on $[-\pi,\pi]$, where

$$\sigma_n(x) = \frac{s_0(x) + \cdots + s_{n-1}(x)}{n}$$

(i.e. the Fourier Series of f converges uniformly $(C,1)$ to f.)

Proof: Since f is continuous on $[-\pi,\pi]$, it is bounded and uniformly continuous. Thus there exists a number M such that $|f(x)| < M$ for all $x \in [-\pi,\pi]$. By uniform continuity, given any $\epsilon > 0$, there exists a $\delta > 0$ such that $|x - y| \leq \delta$ in $[-\pi,\pi]$ implies $|f(x) - f(y)| \leq \frac{\epsilon}{2}$. We will show that for some N, $n \geq N$ implies $|f(x) - \sigma_n(x)| < \epsilon$ for every $x \in [-\pi,\pi]$.

Now

$$|\sigma_n(x) - f(x)| \overset{37.6}{=} \left| \frac{2}{\pi} \int_0^\pi \frac{f(x+t) + f(x-t)}{2} K_n(t)dt - f(x) \right|$$

$$\overset{37.7}{=} \left| \frac{2}{\pi} \int_0^\pi \frac{f(x+t) + f(x-t)}{2} K_n(t)dt - f(x)\frac{2}{\pi}\int_0^\pi K_n(t)dt \right|$$

$$\leq \left| \frac{2}{\pi} \int_0^\delta \frac{f(x+t) + f(x-t) - 2f(x)}{2} K_n(t)dt \right|$$

$$+ \left| \frac{2}{\pi} \int_\delta^\pi \frac{f(x+t) + f(x-t) - 2f(x)}{2} K_n(t)dt \right|$$

$$\leq \frac{\epsilon}{2}\frac{2}{\pi}\int_0^\delta K_n(t)dt$$

$$+ \frac{2}{4n\pi \sin^2(\delta/2)}\int_\delta^\pi [|f(x+t)| + |f(x-t)| + 2|f(x)|]dt$$

$$< \frac{\epsilon}{2} + \frac{1}{2n\pi \sin^2(\delta/2)}\int_\delta^\pi 4M dt.$$

The second term is clearly $< \frac{\epsilon}{2}$ for n large enough, so the result follows. $\square$

We are finally in a position to prove the main theorem of this section.

37.9 Theorem: Let $f \in \mathcal{L}^2[-\pi,\pi]$ and s_n be the nth partial sum of the Fourier Series for f. Then s_n converges to f in the $\mathcal{L}^2$ norm. That is, given any $\epsilon > 0$, there exists an N such that $n > N$ implies $\|s_n - f\|_2 < \epsilon$.

Proof: Given $\epsilon > 0$, by Theorem 33.14 there exists a continuous function g such that $\|f - g\|_2 < \frac{\epsilon}{2}$. By the previous theorem, σ_n (of the Fourier series for g) converges to g uniformly on $[-\pi,\pi]$, so by Exercise 34.29 there is an N such that $\|\sigma_N - g\|_2 < \frac{\epsilon}{2}$.

By the triangle inequality, $\|f - \sigma_N\|_2 < \epsilon$. Since σ_N is a trigonometric polynomial, Theorem 36.1 says that $\|f - s_N\|_2 \leqslant \|f - \sigma_N\|_2 < \epsilon$, and by Corollary 36.5 $\|f - s_n\|_2 < \epsilon$ for $n > N$. □

37.10 Corollary (Parseval's Formula): If $f \in \mathcal{L}^2([-\pi, \pi])$, then

$$\frac{1}{\pi}\left\|f\right\|_2^2 = \frac{a_0}{2} + \sum_{k=1}^{\infty} (a_k^2 + b_k^2).$$

Proof: Exercise 38.20.

We have now seen that the study of Fourier Series is rather nice when viewed in the $\mathcal{L}^2$ setting. We have seen that $\mathcal{L}^2$ is a complete inner product space in which one can integrate convergent series term by term. The continuous functions are dense in $\mathcal{L}^2$ (in the $\mathcal{L}^2$ norm), and the Fourier Series of each $\mathcal{L}^2$ function converges to that function (in the $\mathcal{L}^2$ norm). In most applications, however, one is interested in pointwise convergence. We have seen that $\mathcal{L}^2$ convergence does not imply pointwise convergence. The study of pointwise convergence of Fourier Series does not yield such elegant results as those for $\mathcal{L}^2$. In fact, no nontrivial necessary and sufficient conditions for convergence at a point are known. In the next Chapter we will produce several of the well-known sufficient conditions for pointwise convergence of Fourier Series.

38. Exercises

38.1 (a) Let f be an even function in $\mathcal{L}^2([-\pi, \pi])$ – that is, $f(-x) = f(x)$ for all $x \in [-\pi, \pi]$. Show that in the Fourier series for f, $b_k = 0$ for all k.

   (b) If f is odd – that is, $f(-x) = -f(x)$ – show that $a_k = 0$ for all k.

38.2 If $f \in \mathcal{L}([-\pi, \pi])$, why is $f(x)\cos nx \in \mathcal{L}([-\pi, \pi])$?

38.3 Find the Fourier series for the following functions.

   (a) $f(x) = x$ on $[-\pi, \pi]$.

   (b) $f(x) = 3\cos 2x + 4\sin 5x + 2$.

   (c) $f(x) = 3$.

   (d) $f(x) = \begin{cases} 0 \text{ if } -\pi \leqslant x < 0 \\ x \text{ if } 0 \leqslant x \leqslant \pi. \end{cases}$

   (e) $f(x) = \begin{cases} 0 \text{ if } -\pi \leqslant x < 0 \\ \sin x \text{ if } 0 \leqslant x \leqslant \pi. \end{cases}$

38.4 On graph paper, carefully draw graphs of the first four or five terms of the Fourier series for one or two of the functions in Exercise 38.3. On the same graph, draw the sum of the first two terms, of the first three terms, and so on.

38.5 Prove Corollary 36.6 in detail.

38.6 (a) Use Bessel's Inequality and the Fourier series of Example 35.3 to show that

$$\sum_{k=1}^{\infty} 1/(2k-1)^2 \leqslant \pi^2/8.$$

(b) Use Example 35.5 to get an upper bound on $\sum_{k=1}^{\infty} 1/(2k-1)^4$.

38.7 (a) Prove that $\sum_{n=0}^{\infty} (-1)^n$ converges to $\frac{1}{2}$ $(C,1)$.

(b) Does the series $\sum_{n=1}^{\infty} n(-1)^n$ converge $(C,1)$?

38.8 If $0 \leqslant a_n \leqslant b_n$ for all n and $\sum_{n=1}^{\infty} b_n$ converges $(C,1)$, show that $\sum_{n=1}^{\infty} a_n$ also converges $(C,1)$.

38.9 (a) If $a_n \geqslant 0$ for all n, show that $\sigma_n \leqslant \sigma_{n+1}$ for all n.

(b) If $a_n \geqslant 0$ for all n, and $\sum_{n=1}^{\infty} a_n = \infty$, show that $\lim_{n \to \infty} \sigma_n = \infty$. This problem shows that a series with non-negative terms converges to a real number if and only if it converges $(C,1)$ to that number.

38.10 Alter the proof of Lemma 37.4 to show that for $\theta \neq n\pi$, n an integer,

$$\sum_{k=1}^{n} \sin k\theta = \frac{1}{2\sin\theta} (1 + \cos\theta - \cos n\theta - \cos(n+1)\theta).$$

38.11 (a) Show that $\sum_{k=0}^{\infty} \cos kx$ diverges.

(b) Use Lemma 37.4 to show that the partial sums of the series $\sum_{k=0}^{\infty} \cos kx$ are bounded, for $x \neq 2n\pi$, n an integer.

(c) Use Lemma 37.4 to show that $\sum_{k=0}^{\infty} \cos kx$ converges $(C,1)$ to 0. (Hint: calculate σ_n. See the proof of Corollary 37.6, or use Exercise 38.10.)

38.12 Show that if $\sum_{k=1}^{\infty} a_k$ converges, then

$$\lim_{N \to \infty} \sum_{k=1}^{N} \left(\frac{k-1}{N} \right) a_k = 0.$$

38.13 Draw the periodic extensions of each of the following functions to all of $\mathbb{R}$. Is the resulting function continuous?
(a) $f(x) = x^2$ on $[-\pi, \pi)$.
(b) $f(x) = |x|$ on $[-\pi, \pi)$.
(c) $f(x) = \begin{cases} 0 \text{ if } -\pi \leqslant x < 0 \\ x \text{ if } 0 \leqslant x < \pi. \end{cases}$

38.14 If $f \in \mathcal{L}([-\pi,\pi])$, extended periodically to $\Re$, show that

$$\int_{x-\pi}^{x+\pi} f(t)\,dt = \int_{-\pi}^{\pi} f(t)\,dt.$$

(Hint: draw a picture.)

38.15 Prove Lemma 37.7 in detail.

38.16 Prove the following analogue to Lemma 37.7:
   (a) $\frac{2}{\pi}\int_0^\pi D_n(t)\,dt = 1$
   (b) For $0 < \delta \leqslant |t| \leqslant \pi$, $|D_n(t)| \leqslant 1/2 \,|\sin(\delta/2)\,|$.

38.17 Using Exercise 38.16 and Lemma 37.5 in place of Lemma 37.7 and Corollary 37.6, mimic the proof of Theorem 37.8 and attempt to prove that s_n converges uniformly to f for f continuous on $[-\pi,\pi]$. Does the proof work?

38.18 (a) Show that if f is continuous on $[-\pi,\pi]$ and its Fourier coefficients are all zero, then $f \equiv 0$. (Hint: Theorem 37.8.)
   (b) Show that if f and g are continuous on $[-\pi,\pi]$ and the Fourier coefficients of f and g are identical, then $f(x) = g(x)$ for all x.
   (c) If f and $g \in \mathcal{L}^2([-\pi,\pi])$ have identical Fourier coefficients, what can you conclude? Prove your assertion.

38.19 Recall that the Taylor's series for $\cos x$ (or $\sin x$), expanded about $x = 0$, is a power series with infinite radius of convergence, and that the series converges uniformly to $\cos x$ (or to $\sin x$) on $[-\pi,\pi]$.
   (a) Show that any trigonometric polynomial can be uniformly approximated by a polynomial on $[-\pi,\pi]$); that is, given a trigonometric polynomial t and any $\epsilon > 0$, there is a polynomial p such that $|t(x) - p(x)| < \epsilon$ for all $x \in [-\pi,\pi]$.
   (b) Prove the celebrated Weierstrass Approximation Theorem: every continuous function on a bounded closed interval can be uniformly approximated by a polynomial. (Hint: first justify restriction to the closed interval $[-\pi,\pi]$. Then approximate with a trigonometric polynomial, and use part (a).)
   (c) Describe how you would construct a polynomial p such that

$$||x| - p(x)| < .01 \text{ for all } x \in [-\pi,\pi].$$

38.20 Prove Parseval's Formula: If $f \in \mathcal{L}^2([-\pi,\pi])$, then

$$\frac{1}{\pi}\left\|f\right\|_2^2 = \frac{a_0^2}{2} + \sum_{k=1}^{\infty} (a_k^2 + b_k^2).$$

Note the similarity to Bessel's Inequality. (Hint: Theorem 37.9 says that $\lim \| s_n - f\|_2^2 = 0$. Compute $\| s_n - f\|_2^2$.)

38.21 (a) Show, using Parseval's Formula, that $\sum\limits_{k=1}^{\infty} 1/(2k - 1)^2 = \pi^2/8$. (Example 35.3.)

(b) Find the sum of $\sum_{k=1}^{\infty} 1/(2k-1)^4$. (Example 35.5)

(c) Find $\sum_{k=1}^{\infty} 1/k^2$. (Let $f(x) = x$ on $[-\pi,\pi]$.)

38.22 If $f, g \in \mathcal{L}^2([-\pi,\pi])$, with Fourier series

$$\frac{a_0^2}{2} + \sum_{n=1}^{\infty} a_n \cos nx + b_n \sin nx, \qquad \frac{\alpha_0^2}{2} + \sum_{n=1}^{\infty} \alpha_n \cos nx + \beta_n \sin nx,$$

respectively, show that

$$\frac{1}{\pi}(f \cdot g) = \frac{a_0 \alpha_0}{2} + \sum_{n=1}^{\infty} (a_n \alpha_n + b_n \beta_n).$$

(Hint: Consider $\| f + g \|_2^2$, $\| f - g \|_2^2$ and Parseval's Formula.)

38.23 Compute the Fourier series of $f(x) = e^x$ on $[-\pi,\pi]$ and of $g(x) = x$ on $[-\pi,\pi]$, and use Exercise 38.22 to obtain the sum of $\sum_{n=0}^{\infty} 1/(1 + n^2)$.

38.24 If $f \in \mathcal{L}^2([-\pi,\pi])$ and $x \in [-\pi,\pi]$, show that $\int_{-\pi}^{x} f(t)\,dt$ is equal to the sum of the series obtained by integrating the Fourier series of f term by term. (Hint: let $g(t) = 1$ if $-\pi \le t \le x$ and $g(t) = 0$ if $x < t$. Use Exercise 38.22.)

38.25 Use Example 35.3 and Exercise 38.24 to show that

$$|x| = \pi - (4/\pi) \sum_{n=1}^{\infty} \frac{\cos(2n-1)x + 1}{(2n-1)^2}, \text{ for } x \in [-\pi,\pi].$$

38.26 (a) Use Example 35.5 and Exercise 38.24 to show that for $\pi \ge x \ge 0$,

$$\pi x - x^2 = (8/\pi) \sum_{n=1}^{\infty} \frac{\sin(2n-1)x}{(2n-1)^3}.$$

(b) Let $x = \pi/2$ to obtain the sum of

$$\sum_{n=1}^{\infty} \frac{(-1)^n}{(2n-1)^3}.$$

38.27 If f is continuously differentiable on $[-\pi,\pi]$, show that the Fourier series of f' can be obtained from that of f by differentiating term by term. (Hint: integrate by parts.)

38.28 Show that if $f(x_0^+) = \lim_{x \to x_0^+} f(x)$ and $f(x_0^-) = \lim_{x \to x_0^-} f(x)$ exist, for $x_0 \in (-\pi,\pi)$ and $f \in \mathcal{L}([-\pi,\pi])$, then $\lim_{n \to \infty} \sigma_n(x_0) = \frac{1}{2}[f(x_0^+) + f(x_0^-)]$.

38.29 Suppose that $\sum\limits_{n=1}^{\infty} a_n$ and $\sum\limits_{n=1}^{\infty} b_n$ are absolutely convergent series. Show that

$$\frac{a_0}{2} + \sum_{k=1}^{\infty} (a_k \cos kx + b_k \sin kx)$$

converges absolutely and uniformly on $[-\pi, \pi]$, and its sum is continuous. (Hint: Weierstrass M test.)

38.30 Show that if f is continuous on $[a,b]$ and the Fourier series of f converges at x, then it converges to the value $f(x)$ at x. (Hint: Theorem 37.8.)

38.31 Conclude from the two previous exercises that the Fourier series for $|x|$ converges to $|x|$ for all $x \in [-\pi, \pi]$.

38.32 Prove that if

$$\frac{a_0}{2} + \sum_{n=1}^{\infty} (a_n \cos nx + \sin nx)$$

is the Fourier Series of a continuously differentiable function, then $\sum\limits_{n=1}^{\infty} n a_n$ and $\sum\limits_{n=1}^{\infty} n b_n$ converge (See Exercise 38.27).

CHAPTER 9

Pointwise Convergence of Fourier Series

39. An Application: Vibrating Strings

In Chapter 8 we studied the rather abstract but elegant $\mathcal{L}^2$ theory of Fourier series and obtained a result about $(C,1)$ summability along the way. Of course, Fourier series did not arise as a neat way of expanding $\mathcal{L}^2$ functions, but rather as solutions to a certain class of physical problems. We will examine, rather informally, one such problem: vibrating strings.

Suppose that a string is stretched under tension along the x-axis with fixed endpoints at $x = 0$ and $x = \pi$. If the string is stretched out of its original shape and released, it will tend to vibrate; these vibrations will (under certain conditions) be transmitted to our ears via pressure waves in the air, and we will hear a musical tone.

In order to analyze the situation mathematically, we introduce a function of two variables $y = F(x, t)$, equal to the vertical displacement of the string at position $x \in [0, \pi]$ and time t. Theoretical mechanics gives us a partial differential equation satisfied by this function under certain conditions (for example, displacements must be small, there must be no friction, etc.). The equation is

$$(*) \qquad \frac{\partial^2 y}{\partial t^2} = c^2 \frac{\partial^2 y}{\partial x^2}$$

for a constant c which depends on the physical properties of the string. Rather than attempt to solve this equation in general, we will examine some specific solutions under rather stringent assumptions.

Suppose for the moment that at time $t = 0$ the string is deformed into the shape of a sine curve—

$$F(x,0) = b \sin x, \quad x \in [0,\pi].$$

Assume also that the string is at rest in this position at $t = 0$ so that $F_t(x,0) = 0$ for all $x \in [0,\pi]$. Then if the string is released from this

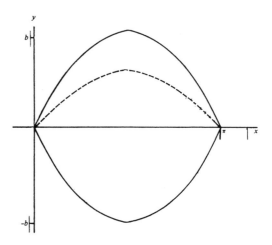

position at $t = 0$, it can be shown that it vibrates sinusoidally within the envelope $y = \pm b \sin x$. In particular,

$$F(x,t) = b \sin x \cos ct$$

for $x \in [0,\pi]$, $t \geqslant 0$. It is easily verified that this function satisfies the basic differential equation (*) as well as the conditions

$$F(x,0) = b \sin x, \quad F_t(x,0) = 0.$$

Notice that since $\cos ct$ goes through an entire cycle (from $+1$ to -1) whenever t changes by $2\pi/c$, the frequency of the vibration (number of beats per unit time) will be $c/2\pi$.

Now suppose instead that our original deformation was of the form

$$F(x,0) = b_n \sin nx,$$

a sine curve with n "humps," in $[0,\pi]$, and suppose again that our initial velocity at each point is 0 so that $F_t(x,0) = 0$. Then a solution in this case is

$$F(x,t) = b_n \sin nx \cos nct.$$

(Verify!) That is, we have a sinusoidal vibration at each x which is bounded by

$$\pm b_n \sin nx.$$

This vibration has $n + 1$ nodes (points at rest) and its frequency is $nc/2\pi$.

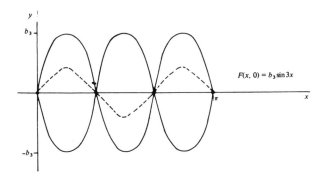

$$F(x, 0) = b_3 \sin 3x$$

The sequence of frequencies

$$\frac{c}{2\pi}, \frac{2c}{2\pi}, \ldots$$

generated above is called the *harmonic* or *overtone series* for the string. The first frequency, $c/2\pi$, is called the *fundamental* or *first harmonic*. The second, $2c/2\pi$, is called the *first overtone* or *second harmonic*, and so forth. We have stated that if the string is initially deformed at zero velocity into the shape $y = b_n \sin nx$, then the string will vibrate purely with the frequency of the nth harmonic.

The question now is what happens when the string is initially deformed, again with zero initial velocity, into an arbitrary shape $F(x,0) = f(x)$. (Of course f cannot really be arbitrary; it must still obey the special assumptions such as small amplitude which were made in deriving the basic differential equation. Also $f(0)$ and $f(\pi)$ should equal zero and the function must be continuous.) The solution could be obtained if we were able to write f in terms of a sine series:

$$f(x) = \sum_{n=1}^{\infty} b_n \sin nx.$$

Since we know all about the vibrations generated by $b_n \sin nx$, it would be reasonable to hope that we could superimpose the vibrations resulting from each $b_n \sin nx$; that is

$$F(x,t) = \sum_{n=1}^{\infty} b_n \sin nx \cos nct.$$

Indeed, formal differentiation term by term (assuming it is justified) shows this is a solution. The meaning of this series solution is that the string vibrates simultaneously with all of the frequencies $nc/2\pi$, for $n = 1, 2, \cdots$. That is, all harmonics are present, each with different amplitude b_n (some of which may be zero). The fundamental frequency usually determines the pitch we hear, and the amplitude of the overtone frequencies determines the quality or *timbre* of the note.

Without discussion let us assert that similar results are obtained if we do not assume that the initial velocity of the string is zero.

Our problem, then, has been reduced to the question of whether a given f can be represented by a sine series

$$\sum_{n=1}^{\infty} b_n \sin nx.$$

In other applications (where f is not necessarily 0 at $x = 0$ and $x = \pi$), we must consider representations of f by *trigonometric* series

$$f(x) = \frac{a_0}{2} + \sum_{k=1}^{\infty} (a_k \cos kx + b_k \sin kx).$$

If we assume that f has such a representation, and that term by term integration is legitimate (as for $\mathcal{L}^2$ functions), then we find, as in the proof of Theorem 36.6, that a_k and b_k are the Fourier coefficients of f. In other words, if we are looking for a representation by trigonometric series, the only reasonable candidate is the Fourier series.

Now, we know that every $f \in \mathcal{L}[-\pi, \pi]$ has a Fourier series. The question is, for which Lebesgue summable functions f and which points x does this Fourier series converge to $f(x)$. In the following sections we will give some partial answers to this question. It is gratifying that most of the nice functions we meet in daily life do have Fourier series which converge to the function at least for most points.

40. Some Bad Examples and Good Theorems

Since we can change a function on a set of measure zero without changing its Fourier series, it is obvious that Fourier series do not always converge to $f(x)$ for each x. If we restrict ourselves to continuous functions, however, we cannot alter a function on a set of measure zero without destroying continuity. It is tempting to believe, therefore, that the Fourier series of a continuous function f converges to $f(x)$ for every x. The following outline of an example (due to Fejer) shows that this is not the case.

40.1 Example: Given a positive integer n and $x \in [-\pi, \pi]$, let

$$s_n = \frac{\cos nx}{n} + \frac{\cos(n+1)x}{n-1} + \cdots + \frac{\cos(2n-1)x}{1}$$

$$-\frac{\cos(2n+1)x}{1} - \frac{\cos(2n+2)x}{2} - \cdots - \frac{\cos(3n)x}{n}.$$

Then it can be shown (though not easily) that $\{s_n(x)\}$ is uniformly bounded, i.e. there is a positive number M such that $|s_n(x)| < M$ for all n, x.

Now for $k = 1, 2, \cdots$, let $n_k = 2^{k^2}$, and consider the series

$$\sum_{k=1}^{\infty} \frac{1}{k^2} s_{n_k}(x).$$

Since

$$\left| \frac{1}{k^2} s_{n_k}(x) \right| < M \frac{1}{k^2},$$

this series converges uniformly by the Weierstrass M-test to a continuous function f.

Hence

$$f(x) = 1\left(\frac{\cos 2x}{2} + \frac{\cos 3x}{1} - \frac{\cos 5x}{1} - \frac{\cos 6x}{2} \right)$$

$$+ \frac{1}{4}\left(\frac{\cos 16x}{16} + \frac{\cos 17x}{15} + \cdots + \frac{\cos 31x}{1} - \frac{\cos 33x}{1} \right.$$

$$\left. - \frac{\cos 34x}{2} - \cdots - \frac{\cos 48x}{16} \right) + \frac{1}{9}\left(\frac{\cos 256x}{256} + \cdots \right)$$

$$+ \cdots$$

Let $g_i(x)$ be the ith term of the series obtained by dropping the parentheses; that is,

$$\sum_{i=1}^{\infty} g_i(x) = \frac{\cos 2x}{2} + \frac{\cos 3x}{1} - \frac{\cos 5x}{1} - \frac{\cos 6x}{2} + \frac{1}{4}\frac{\cos 16x}{16} + \frac{1}{4}\frac{\cos 17x}{15} + \cdots$$

Then this cosine series is the Fourier series for f (e.g. for $p > 0$,

$$a_p = \frac{1}{\pi}\int_{-\pi}^{\pi}\left(\sum_{k=1}^{\infty}\frac{1}{k^2}s_{n_k}(x)\right)\cos px\,dx = \frac{1}{\pi}\sum_{k=1}^{\infty}\frac{1}{k^2}\int_{-\pi}^{\pi}s_{n_k}(x)\cos px\,dx.)$$

However, the series $\sum_{i=0}^{\infty} g_i(0)$ diverges. We show it it not a Cauchy series by noting that given N, there is an $N_1 \geqslant N$ such that for some k,

$$g_{N_1}(0) = \frac{1}{k^2}\cdot\frac{1}{n_k},$$

$$g_{N_1+1}(0) = \frac{1}{k^2}\cdot\frac{1}{n_k-1},\cdots,$$

$$g_{N_1+n_k-1}(0) = \frac{1}{k^2}\cdot\frac{1}{1}.$$

Therefore, if s_n is the nth partial sum of $\sum_{i=1}^{\infty} g_i(0)$,

$$(s_{N_1+n_k-1} - s_{N_1})$$

$$= \frac{1}{k^2}\left(\frac{1}{n_k} + \frac{1}{n_k-1} + \frac{1}{n_k-2} + \cdots + \frac{1}{2} + 1\right) \geqslant \frac{1}{k^2}\int_{1}^{n_k}\frac{1}{t}dt$$

$$= \frac{1}{k^2}\log n_k = \frac{1}{k^2}\log 2^{k^2} = \frac{1}{k^2}\cdot k^2\log 2 = \log 2.$$

Therefore, the Fourier series of f fails to converge at 0, and thus does not converge to $f(0)$.

There are even worse examples involving continuous functions. If E is any set of measure zero in $[-\pi,\pi]$ then there exists a continuous function whose Fourier series diverges (unboundedly) on E (see Buzdalin, V.V.; "Unboundedly Diverging Trigonometric Fourier Series of Continuous Functions," Math. Notes 7 (1970), pgs 5-12.)

For many years mathematicians asked whether the Fourier series of an arbitrary continuous function had to converge at any point. Finally, in

1966, Lenart Carleson showed that the Fourier series of every $\mathcal{L}^2$ function (hence of every continuous function) converges pointwise a.e. [see Carleson, Lenart; "On Convergence and Growth of Partial Sums of Fourier Series," Acta Mathematica 116 (1966), pgs 135-157]. It should be noted that there exist summable (necessarily non-$\mathcal{L}^2$) functions whose Fourier series diverge *everywhere*.

Thus, theoretically, things are not too bad. However, to apply Fourier series to physical problems, one often needs to know whether a *particular* function has a convergent Fourier series at a *particular* point.

We now present some sufficient conditions under which the Fourier series of f converges to $f(x)$ at x. At first the hypotheses may seem exceedingly strong, but one must remember that many applied problems involve functions which have nice properties. The obvious property to try (since continuity failed) is differentiability.

40.2 Theorem: Let $f \in \mathcal{L}^2[-\pi,\pi]$ and let f be differentiable at $x_0 \in [-\pi,\pi]$. Then the Fourier series for f converges to $f(x_0)$ at x_0.

Proof: From Lemma 37.5, the nth Fourier partial sum is

$$s_n(x_0) = \frac{1}{\pi} \int_0^\pi [f(x_0 + t) + f(x_0 - t)] D_n(t) dt.$$

(Recall that we assume f is extended periodically when necessary as in section 37.) Also, for the particular case $f(x) \equiv 1$, we obtain

$$1 = \frac{2}{\pi} \int_0^\pi D_n(t) dt \qquad \text{for } n = 1,2,3, \cdots.$$

Thus for our general f, differentiable at x_0,

$$s_n(x_0) - f(x_0) = \frac{1}{\pi} \int_0^\pi [f(x_0 + t) + f(x_0 - t) - 2f(x_0)] D_n(t) dt.$$

Now

$$h(t) = \frac{f(x_0 + t) + f(x_0 - t) - 2f(x_0)}{2 \sin \frac{1}{2} t}$$

can be defined so as to be continuous at $t = 0$. (Compute

$$\lim_{t \to 0} \frac{f(x_0 + t) - f(x_0) + f(x_0 - t) - f(x_0)}{t} \cdot \frac{t}{2 \sin \frac{1}{2} t}.$$

Note that this is where the differentiability of f at x_0 is used. See Exercise 42.4.) Therefore, $h \in \mathcal{L}^2[0,\pi]$ (see Exercise 42.5). Thus,

$$s_n(x_0) - f(x_0) = \frac{1}{\pi} \int_0^\pi [f(x_0 + t) + f(x_0 - t) + 2f(x_0)]D_n(t)dt$$

$$= \frac{1}{\pi} \int_0^\pi h(t)\sin\left(n + \frac{1}{2}\right)t\,dt.$$

Thus by the Riemann-Lebesgue Lemma (Corollary 36.4), $s_n(x_0) - f(x_0) \to 0$ as $n \to \infty$. Note that the Riemann-Lebesgue Lemma holds in this case by Exercise 42.6. ☐

We now present a condition on $f(x)$ which will guarantee uniform convergence of the Fourier series for F. We first remind the reader that a function is said to be C^2 on an interval $[a,b]$ if its second derivative is continuous throughout $[a,b]$.

40.3 Theorem: Let f be C^2 on $[-\pi, \pi]$. Then the Fourier series of f converges uniformly to f on $[-\pi, \pi]$.

Proof: The previous theorem guarantees pointwise convergence of the Fourier series to $f(x)$ at each x. To show uniform convergence recall

$$a_n = \frac{1}{\pi} \int_{-\pi}^\pi f(x)\cos nx\,dx \quad \text{and} \quad b_n = \frac{1}{\pi} \int_{-\pi}^\pi f(x)\sin nx\,dx$$

which in this case can be taken to be Riemann integrals. Integrate by parts twice to show that there exists a positive number M such that $|a_n|$ and $|b_n|$ are both $< M/n^2$. By the Weierstrass M-test, the Fourier series converges uniformly. (Where is the C^2 hypothesis used?). ☐

In the next section we will present a more general sufficient condition for convergence of Fourier series. However, we again note that many useful functions are differentiable or C^2 so that the rather easy results above provide all the necessary knowledge about convergence of Fourier series in these cases.

41. More Convergence Theorems

In this section we shall examine the local (i.e. near a point) behavior of Fourier series. In order to proceed we need to know that the Riemann-Lebesgue Lemma (Corollary 36.4) holds for $f \in \mathcal{L}[-\pi, \pi]$.

41.1 Lemma (Riemann-Lebesgue): If $f \in \mathcal{L}[-\pi,\pi]$ and $\{a_k\}_{k=0}^{\infty}$ and $\{b_k\}_{k=1}^{\infty}$ are the Fourier coefficients of f, then $\lim\limits_{k\to\infty} a_k = \lim\limits_{k\to\infty} b_k = 0$.

Proof: Given $\epsilon > 0$, there exists a simple function g on $[-\pi,\pi]$ such that $\int_{-\pi}^{\pi} |f(t) - g(t)| dt < \epsilon$. (This follows from the definition of the integral for $f \in \mathcal{L}$.)

Now $g \in \mathcal{L}^2$, so that if A_k, B_k are the Fourier coefficients of g, then $\lim\limits_{k\to\infty} A_k = \lim\limits_{k\to\infty} B_k = 0$ by the Riemann-Lebesgue Lemma for $\mathcal{L}^2$. But for each k,

$$|a_k - A_k| = \left| \frac{1}{\pi} \int_{-\pi}^{\pi} [f(t) - g(t)] \cos kt\, dt \right| \leqslant \frac{1}{\pi} \int_{-\pi}^{\pi} |f(t) - g(t)|\, dt < \epsilon$$

by the choice of g above. Thus $\lim\limits_{k\to\infty} a_k = 0$ and similarly $\lim\limits_{k\to\infty} b_k = 0$. $\qquad\square$

We are now in a position to prove the Riemann Localization Theorem which says that the behavior of the Fourier series at x depends only on the values of f arbitrarily close to x. This is somewhat surprising in view of the definition of the Fourier coefficients of f as integrals over the entire interval $[-\pi,\pi]$. If two functions in $\mathcal{L}[-\pi,\pi]$ agree in a neighborhood of x, then their Fourier series either both diverge at x or both converge to the same value at x, even though the functions may differ greatly outside the neighborhood.

41.2 Theorem (Riemann Localization Theorem): Let $f \in \mathcal{L}[-\pi,\pi]$ and $x \in [-\pi,\pi]$, and let δ be any positive real number. Then

$$\lim_{n\to\infty} \int_{\delta}^{\pi} [f(x+t) + f(x-t)] D_n(t) dt = 0.$$

Proof: Consider

$$g(t) = \begin{cases} 0 \text{ on } [0,\delta) \\ \dfrac{f(x+t)+f(x-t)}{2\sin t/2} \text{ on } [\delta,\pi]. \end{cases}$$

Clearly $g \in \mathcal{L}[-\pi,\pi]$ since $2\sin t/2 \geqslant 2\sin \delta/2 > 0$ on $[\delta,\pi]$. Thus

$$\int_{\delta}^{\pi} [f(x+t) + f(x-t)] D_n(t) dt = \int_{0}^{\pi} g(t) \sin\left(n + \frac{1}{2}\right) t\, dt$$

and

$$\lim_{n \to \infty} \int_{\delta}^{\pi} [f(x + t) + f(x - t)]D_n(t)dt = 0$$

by the Riemann-Lebesgue Lemma and Exercise 42.6. □

41.3 Corollary: If $f \in \mathcal{L}[-\pi, \pi]$ is identically 0 on some open interval containing x, then the Fourier series of f converges to 0 at x.

Proof: Suppose $f \equiv 0$ on $(x - \delta, x + \delta)$. Then the nth partial sum

$$s_n(x) = \frac{1}{\pi} \int_{0}^{\pi} [f(x + t) + f(x - t)]D_n(t)dt$$

$$= \frac{1}{\pi} \int_{\delta}^{\pi} [f(x + t) + f(x - t)]D_n(t)dt.$$

So by the Theorem, $\lim_{n \to \infty} s_n(x) = 0$. □

41.4 Corollary: If $f, g \in \mathcal{L}[-\pi, \pi]$ and $f = g$ on an open interval containing x, then either both Fourier series converge to the same value at x or both diverge at x.

Proof: Use $f - g$ in Corollary 41.3. Note that the Fourier series of $f - g$ equals the Fourier series of f minus the Fourier series of g by the definition of Fourier coefficients and linearity of the integral. □

This local property of Fourier series is in marked contrast to power series results. Recall that if two power series have the same values over an interval, then they are equal on their entire interval of convergence. But Corollary 41.4 says we can alter f drastically outside the interval (hence obtaining greatly different Fourier series) and yet retain the same values inside the interval.

We will now obtain a generalization of a result in section 39 concerning differentiability of f at x. Here we will require only that f have generalized left and right derivatives at x. We first need to define these carefully.

41.5 Definition: Let f be a real valued function with right and left hand limits $f(x^+)$ and $f(x^-)$ respectively at x. Then the *generalized right and left hand derivatives of f at x* are respectively

$$f_R'(x) = \lim_{t \to x^+} \frac{f(t) - f(x^+)}{t - x} \quad \text{and} \quad f_L'(x) = \lim_{t \to x^-} \frac{f(t) - f(x^-)}{t - x}.$$

41.6 Example: Let

$$f(x) = \begin{cases} x^2 + 1 \text{ on } [-\pi, 0) \\ 10 \text{ at } 0 \\ x - 2 \text{ on } (0, \pi). \end{cases}$$

Then $f(0^+) = -2$, $f(0^-) = 1$, $f_R'(0) = 1$ and $f_L'(0) = 0$ (Verify!). Notice that f has no ordinary right and left hand derivatives at 0 since f is not continuous from either the right or left at 0.

41.7 Theorem: Let $f \in \mathcal{L}[-\pi, \pi]$ and $f_R'(x)$ and $f_L'(x)$ both exist. Then the Fourier series for f converges to

$$\frac{f(x^+) + f(x^-)}{2} \text{ at } x.$$

Proof: This is almost exactly like the proof of Theorem 40.2. See Exercise 42.9. □

As an example of the above theorem, recall from Example 35.3 that the Fourier series for

$$f(x) = \begin{cases} 0 \text{ on } [-\pi, 0) \\ 1 \text{ on } [0, \pi] \end{cases}$$

has the value

$$\frac{1}{2} = \frac{f(0^+) + f(0^-)}{2} \text{ at } 0.$$

Note that the above theorem truly generalizes Theorem 39.2 since if f is differentiable, it clearly has f_R' and f_L' and thus its Fourier series converges to

$$\frac{f(x) + f(x)}{2} = f(x) \text{ at } x.$$

42. Exercises

42.1 In reference to Carleson's result (section 39), show that if the Fourier series of $f \in L^2([-\pi,\pi])$ converges pointwise a.e., then it converges pointwise a.e. to f. (Hint: Show, as in the proof of Theorem 33.11, that a subsequence of the sequence of Fourier partial sums converges pointwise a.e. to f.)

42.2 Consider the Fourier series of Example 35.3.
(a) Show that $\pi/4 = 1 - 1/3 + 1/5 - 1/7 + \dots$ (let $x = \pi/2$).
(b) Let $x = \pi/4$ to obtain another series expression for $\pi/4$.

42.3 Using the Fourier series for x^2 on $[-\pi,\pi]$, find the sum of the series

$$\sum_{k=1}^{\infty} \frac{(-1)^{k+1}}{k^2}.$$

See what other numerical series you can sum with this Fourier series.

42.4 Prove that

$$h(t) = \frac{f(x_0 + t) + f(x_0 - t) - 2f(x_0)}{2\sin\frac{1}{2}t},$$

for f differentiable at $x_0 \in [-\pi,\pi]$, can be defined so as to be continuous at $t = 0$.

42.5 If $h(t) = f(t)/g(t)$ for $t \neq t_0$, $t_0 \in [a,b]$, $g(t) \neq 0$ except at $t = t_0$, $f \in L^2([a,b])$, g continuous, and $\lim_{t \to t_0} f(t)/g(t) = c \in R$, then $h \in L^2([a,b])$.

42.6 If $h \in L^2([-\pi,\pi])$, show that $\lim_{n \to \infty} \int_0^\pi h(t)\sin(n + \frac{1}{2})t\,dt = 0$.

42.7 Fill in the details of the proof of Theorem 40.3.

42.8 Given $f \in L([-\pi,\pi])$ and $\epsilon > 0$, show that there is a simple function g on $[-\pi,\pi]$ such that $\int_{-\pi}^\pi |f(t) - g(t)|\,dt < \epsilon$. (Hint: apply the definition of the integral for f^+, f^-.)

42.9 Prove Theorem 41.7.

42.10 Look at the Fourier series of Example 35.3 and 35.5 to verify Theorem 41.7 at points of discontinuity.

42.11 (a) For $x \in [0,\pi]$, find the Fourier series for the function $\chi_{[0,x)}$.
(b) Obtain from (a) the expression, for $x \in [0,\pi]$,

$$\frac{\pi - x}{2} = \sum_{n=1}^{\infty} \frac{\sin nx}{n}.$$

(c) Let $x = \pi/2$ in (b) to obtain an interesting series.

42.12 (a) Show that if a trigonometric series

$$\frac{a_0}{2} + \sum_{n=1}^{\infty} (a_n \cos nx + b_n \sin nx)$$

converges absolutely at c, then it converges absolutely at $-c$.

(b) Show that if a trigonometric series converges absolutely at c_1 and c_2, then it converges absolutely at $c_1 + c_2$.

(c) Show that if a trigonometric series converges absolutely on any open interval, then it converges absolutely everywhere.

Appendix

Logic and Sets

The connectives and quantifiers of logic, as commonly used in mathematics, are closely related to operations on sets. For example, the definition $A \cap B = \{x \mid x \in A \text{ and } x \in B\}$ expresses the relationship between intersection and the logical connective "*and*." Of course $A \cup B = \{x \mid x \in A \text{ or } x \in B\}$, and $A \backslash B = \{x \mid x \in A \text{ and } x \notin B\}$. This last set is called the *relative complement* of B in A.

Given a sequence of sets $\{A_n\}_{n=0}^{\infty}$, one can form the union or the intersection of the sequence by making use of the logical connectives $\exists$ ("there exists") and $\forall$ ("for all"):

$$\bigcup_{n=0}^{\infty} A_n = \{x \mid (\exists n)(x \in A_n)\}$$

$$\bigcap_{n=0}^{\infty} A_n = \{x \mid (\forall n)(x \in A_n)\}.$$

The reader can verify such identities as the following:

$$B \backslash (\bigcup_{n=0}^{\infty} A_n) = \bigcap_{n=0}^{\infty} (B \backslash A_n)$$

$$B \backslash (\bigcap_{n=0}^{\infty} A_n) = \bigcup_{n=0}^{\infty} (B \backslash A_n)$$

$$B \cup (\bigcap_{n=0}^{\infty} A_n) = \bigcap_{n=0}^{\infty} (B \cup A_n).$$

Open and Closed Sets

A set A of real numbers is *open* (in $\mathcal{R}$) if for each $x \in A$, there is an open interval (a,b) such that $x \in (a,b)$ and $(a,b) \subset A$. A set B is *closed* if $\mathcal{R} \backslash B$ is open.

Sometimes it is desirable to focus attention on a subset X of $\mathcal{R}$ and to speak of subsets $A \subset X$ as being open or closed *relative to X*. A set $A \subset X$ is *open in X* (or "in the relative topology of X") if there is a set U, open in $\mathcal{R}$, such that $A = U \cap X$. (Example: $[0,\frac{1}{2})$ is open in the relative topology of $[0,1]$ since $[0,\frac{1}{2}) = (-1,\frac{1}{2}) \cap [0,1]$.) The reader can verify that a set $A \subset X$ is open in X if and only if for every $x \in A$ there is an open interval (a,b) such that $x \in (a,b)$ and $(a,b) \cap X \subset A$.

Given $x \in X$, a *neighborhood of x* in X is a set $A \subset X$, open in X and containing x.

Bounded Sets of Real Numbers

A real number u is an upper bound of a set of real numbers X if $(\forall x \in X)[u \geqslant x]$. A number ℓ is a lower bound if $(\forall x \in X)[\ell \leqslant x]$.

It is useful to allow the two symbols ∞ and $-\infty$ to be used as upper and lower bounds. Of course ∞ is considered to be larger than any real number and $-\infty$ is considered to be smaller than any real number. Thus ∞ is an upper bound for any set of real numbers and $-\infty$ is lower bound for any set of real numbers. Since the null set $\emptyset$ has no elements, it has $-\infty$ as an upper bound and ∞ as a lower bound.

It is a fundamental property of the real numbers (The Completeness Axiom) that every set of real numbers has a least upper bound (an upper bound that is less than every other upper bound) and a greatest lower bound (a lower bound that is greater than every other lower bound). It is easy to see that no set can have more than one least upper bound nor more than one greatest lower bound. The least upper bound of a set X is denoted $\text{lub}(X)$ or $\sup(X)$, the latter standing for "supremum." The greatest lower bound of X is denoted $\text{glb}(X)$ or $\inf(X)$, the latter standing for "infimum."

Note that for any non-empty set X of real numbers, $\text{glb}(X) \leqslant \text{lub}(X)$ but that $\text{lub}(\emptyset) = -\infty$ and $\text{glb}(\emptyset) = \infty$.

Countable and Uncountable Sets

Infinite sets are often encountered in mathematics, and it is sometimes useful to distinguish among "different sizes" of infinity. The smallest infinite sets are said to be *countable* or *countably infinite*. They can be put into a one-to-one correspondence with the natural numbers $N = \{0, 1, 2, \cdots\}$. Any infinite set which is not countable is said to be *uncountable*. For example, the integers $Z = \{\cdots -2, -1, 0, 1, 2, \cdots\}$ are countably infinite. A one-to-one correspondence with N which proves this can be given as follows:

$$
\begin{array}{ccccccccc}
Z: & 0 & 1 & -1 & 2 & -2 & 3 & -3 & 4 & \cdots \\
   & \updownarrow & \updownarrow & \updownarrow & \updownarrow & \updownarrow & \updownarrow & \updownarrow & \updownarrow & \\
N: & 0 & 1 & 2 & 3 & 4 & 5 & 6 & 7 & \cdots
\end{array}
$$

This illustration shows the reason for the word "countable" in referring to such sets. The one-to-one correspondence amounts to writing out all the elements of the set in some systematic and non-redundant list, one after the other. They can then be "counted."

A reader who is not familiar with these ideas might suppose that every infinite set is countable. However, the set of all real numbers is uncountable. Indeed any interval of real numbers is uncountable. We will illustrate with the open interval $(0,1)$. The method of proof is known as Cantor's Diagonal Argument.

Suppose that $(0,1)$ were countable. Then its elements could all be listed, one after another. Let us write such a list as $r_1, r_2, r_3, \cdots$. Then each r_i can be represented by an infinite decimal of the form $r_i = 0.a_{i1}a_{i2}a_{i3} \cdots$, where each a_{ij} is an integer and $0 \leqslant a_{ij} \leqslant 9$. (Some elements of $(0,1)$ have two decimal representations: for example, $.2 = .19999 \cdots$. We can avoid some difficulty by agreeing always to choose one of these representations—say the one with the infinite string of 9's.)

Now we create an infinite decimal $r = 0.d_1d_2d_3 \cdots$ as follows: d_1 is chosen to be different from a_{11} (for example, $d_1 = 5$ if $a_{11} \neq 5$ and $d_1 = 6$ if $a_{11} = 5$). This guarantees that $r \neq r_1$ since the two numbers differ at the first decimal place. Now choose d_2 in the same fashion so that $d_2 \neq a_{22}$. Then necessarily $r \neq r_2$. In general, d_i is chosen so that $d_i \neq a_{ii}$, which guarantees $r \neq r_i$. We have created a number $r \in (0,1)$ which is not equal to any number on the list! So our assumption that the list was complete is false, and $(0,1)$ must be uncountable.

The fact that between every two real numbers there is a rational number might lead one to suspect that the set of rational numbers Q is uncountable as well. However, Q is indeed countable, as we shall show. To do so, we will make use of the following results: if $A_0, A_2, A_2, \cdots$ is a countable collection of countable sets, then their union $\bigcup_{i=0}^{\infty} A_i$ is countable. The proof of this result will be discussed in the next paragraph. To prove that Q is countable, let $A_0 = [0,1) \cap Q$, $A_1 = [1,2) \cap Q$, $A_2 = [-1,0) \cap Q$, $A_3 = [2,3) \cap Q$, $A_4 = [-2,-1) \cap Q$, etc. Each A_i is countable, as the following listing shows:

$$[n, n+1) \cap Q = \left\{ n, n + \frac{1}{2}, n + \frac{1}{3}, n + \frac{2}{3}, n + \frac{1}{4}, n + \frac{3}{4}, \right.$$

$$\left. n + \frac{1}{5}, n + \frac{2}{5}, n + \frac{3}{5}, n + \frac{4}{5}, n + \frac{1}{6}, \cdots \right\}.$$

The list is complete, since every possible denominator appears with every meaningful numerator: Therefore $\bigcup_{i=0}^{\infty} A_i = Q$ must be countable.

To demonstrate that a countable union of countable sets is countable, we list the elements of each A_i: $a_{i0}, a_{i1}, a_{i2}, \cdots$. We can then list the elements of $\bigcup_{i=0}^{\infty} A_i$ as follows: $a_{00}, a_{01}, a_{10}, a_{02}, a_{11}, a_{20}, a_{03}, a_{12}, a_{21}, a_{30}, a_{04}, \cdots$. (The rule being followed is to list the elements whose subscripts add up to 0, then those whose subscripts add up to 1, etc.) This seems straightforward once the basic idea is understood. Actually there are a couple of difficulties with the proof. First, there may be duplications in the final list if the A_i are not disjoint. This difficulty is easily overcome by simply eliminating duplications as they arise.

A more fundamental problem arises from the fact that there are infinitely many ways to list each A_i. In specifying a particular listing of the elements in A_i we are in fact making one arbitrary choice (of a particular listing) for each $i = 1, 2, 3, \cdots$. This process of making infinitely many arbitrary choices at the same time cannot be justified on the basis of ordinary mathematical principles. An additional principle, called the *Axiom of Choice*,[1] must be invoked. Although the reader may find this principle trivial, that feeling is probably based on his experience with *finite* situations, where the Axiom of Choice is not needed. Use of

---

[1] The Axiom of Choice states that one can form a set consisting of exactly one element from each set in a given infinite collection of nonempty sets. For a more complete treatment of the Axiom of Choice see *Elements of Modern Algebra* by K.S. Miller (Krieger, 1975).

the Axiom of Choice often leads to results which are obvious and comforting, but sometimes rather paradoxical and startling conclusions can arise. These results, coupled with a fundamental philosophical uneasiness with any principle which allows one to accomplish infinitely many steps all at once, have caused some mathematicians to reject the Axiom of Choice and the conclusions which follow from it. Even mathematicians who are happy with the Axiom of Choice are sufficiently sensitive to note specifically where it is being used in their proofs.

It should be noted that the Axiom of Choice is not needed to prove that Q is countable. (However, the Axiom of Choice is needed in 7.7 to show the existence of a non-measurable set.) The listing of $[n, n + 1) \cap Q$ which we gave has the same form for every n and is specified in advance. Therefore *no* choices need be made in listing $\bigcup_{i=0}^{\infty} A_i$. The reader should verify that the first few terms of the resulting listing, following the rule given earlier, are as follows: $0, \frac{1}{2}, 1, \frac{1}{3}, \frac{3}{2}, -1, \frac{2}{3}, \frac{4}{3}, -\frac{1}{2}, 2, \cdots$.

Real Functions

Let X and Y be two sets of real numbers and let f and g be functions which map X and Y respectively into $\mathfrak{R}$. (It is accepted practice to call X the *domain* of f and Y the *domain* of g). One can define new functions as follows:

$$f + g: X \cap Y \to \mathfrak{R} \quad \text{by} \quad (f + g)(x) = f(x) + g(x)$$

$$fg: X \cap Y \to \mathfrak{R} \quad \text{by} \quad (fg)(x) = f(x)g(x)$$

$$f/g: X \cap \{x \in Y | g(x) \neq 0\} \to \mathfrak{R} \quad \text{by} \quad (f/g)(x) = f(x)/g(x)$$

$$f \cdot g: \{x \in Y | g(x) \in X\} \to \mathfrak{R} \quad \text{by} \quad (f \cdot g)(x) = f(g(x))$$

This last is called "*f composed with g*."

If $f: X \to \mathfrak{R}$, and $A \subset R$, then $\{x \in X | f(x) \in A\}$ is called the "pre-image of A under f" and is denoted $f^{-1}(A)$. This notation does not implicitly assume that f has an inverse (is one-to-one); the notation applies to any function. The reader should verify the following results:

$$f^{-1}(\bigcup_i A_i) = \bigcup_i f^{-1}(A_i)$$

$$f^{-1}(\bigcap_i A_i) = \bigcap_i f^{-1}(A_i)$$

$$f^{-1}(\mathfrak{R}\backslash A) = \mathfrak{R}\backslash f^{-1}(A).$$

If $f: X \to \mathfrak{R}$, and $B \subset X$, then $f(B) = \{f(x) \,|\, x \in B\}$. Note that $f^{-1}(f(B)) \supset B$, with equality holding if f is one-to-one. Similarly, $f(f^{-1}(A)) \subset A$, with equality holding provided f is onto.

A function $f: X \to \mathfrak{R}$ is *bounded* if $f(X) \subset [m,M]$ for some $m,M \in \mathfrak{R}$.

A function $f: X \to \mathfrak{R}$ is *increasing* on $[a,b] \subset X$ if

$$(\forall x,y \in [a,b])(x < y \Rightarrow f(x) \leqslant f(y)),$$

and f is *decreasing* on $[a,b]$ if

$$(\forall x,y \in [a,b])(x < y \Rightarrow f(x) \geqslant f(y)).$$

Also, f is *monotone* on $[a,b]$ if f is either increasing on $[a,b]$ or decreasing on $[a,b]$. A function is *piecewise monotone* on $[a,b]$ if there are finitely many points $a = x_0 < x_1 < \cdots < x_n = b$ such that the function is monotone on $[x_{i-1},x_i]$ for each $i = 1,2,\cdots,n$.

A function $f: X \to R$ is *continuous* at $x \in X$ if

$$(\forall \epsilon > 0)(\exists \delta > 0)(\forall y \in X)(|x - y| < \delta \Rightarrow |f(x) - f(y)| < \epsilon).$$

This is equivalent to saying that for every neighborhood V of $f(x)$ in $\mathfrak{R}$, there exists a neighborhood U of x in X such that $f(U)$ is contained in V. If f is continuous at every point $x \in A \subset X$, then f is said to be *continuous on A*. This is equivalent to saying

$$(\forall V \text{ open in } R)(f^{-1}(V) \text{ is open in } A).$$

A function $f: [a,b] \to \mathfrak{R}$ is *piecewise continuous* on $[a,b]$ if there are finitely many points $a = x_0 < x_1 < \cdots < x_n = b$ such that f is continuous on $[x_{i-1},x_i]$ for each $i = 1,2,\cdots,n$.

It is a major result in the theory of real functions that if f is continuous on $[a,b]$, then f is *uniformly continuous* on $[a,b]$. This means that given $\epsilon > 0$, there is a $\delta > 0$ which works for every $x \in [a,b]$; in other words,

$$(\forall \epsilon > 0)(\exists \delta > 0)(\forall x,y \in [a,b])(|x - y| < \delta \Rightarrow |f(x) - f(y)| < \epsilon).$$

Real Sequences

A sequence $\{a_n\}$ of real numbers has real *limit* L $(\lim_{n \to \infty} a_n = L)$ if

$$(\forall \epsilon > 0)(\exists N)(\forall n > N)(|a_n - L| < \epsilon).$$

Also,

$$\lim_{n \to \infty} a_n = \infty \text{ if } (\forall M)(\exists N)(\forall n > N)(a_n > M),$$

and

$$\lim_{n \to \infty} a_n = -\infty \text{ if } (\forall M)(\exists N)(\forall n > N)(a_n < -M).$$

Not every sequence has a limit of course. A sequence $\{a_n\}$ is *increasing* if $(\forall n)(\forall m)(n < m \Rightarrow a_n \leqslant a_m)$, and $\{a_n\}$ is *decreasing* if $n < m$ implies $a_n \geqslant a_m$. A sequence is *monotone* if it is either increasing or decreasing. Every monotone sequence can be shown to have a limit which is either a real number or $\pm\infty$. This limit is the lub of the set $\{a_n | n = 1, 2, \cdots \}$ if the sequence is increasing, or the glb of the set $\{a_n | n = 1, 2, \cdots \}$ if the sequence is decreasing.

If $\{a_n\}$ is any sequence of real numbers, define

$$b_n = \text{lub}\{a_k | k \geqslant n\} \quad \text{and} \quad c_n = \text{glb}\{a_k | k > n\}$$

for $n = 1, 2, \cdots$. Then it is easy to see that $\{b_n\}$ is a decreasing sequence, $\{c_n\}$ is an increasing sequence, and $b_n \geqslant a_n \geqslant c_n$ for all n. Therefore the monotone sequences $\{b_n\}$, $\{c_n\}$ both have limits. The limit of $\{b_n\}$ is called $\overline{\lim_{n \to \infty}} a_n$ or $\limsup_{n \to \infty} a_n$, and the limit of $\{c_n\}$ is called $\underline{\lim_{n \to \infty}} a_n$ or $\liminf_{n \to \infty} a_n$. The reader can verify that

(1) $\underline{\lim_{n \to \infty}} a_n \leqslant \overline{\lim_{n \to \infty}} a_n$,

(2) $\{a_n\}$ converges if and only if $\underline{\lim_{n \to \infty}} a_n = \overline{\lim_{n \to \infty}} a_n$ (and, in that case, $\lim_{n \to \infty} a_n = \underline{\lim_{n \to \infty}} a_n = \overline{\lim_{n \to \infty}} a_n$),

(3) if $a_n \leqslant d_n$ for all n, then $\underline{\lim_{n \to \infty}} a_n \leqslant \underline{\lim_{n \to \infty}} d_n$ and $\overline{\lim_{n \to \infty}} a_n \leqslant \overline{\lim_{n \to \infty}} d_n$,

(4) $\underline{\lim_{n \to \infty}}(-a_n) = -\overline{\lim_{n \to \infty}} a_n$.

Sequences of Functions

A sequence $\{f_n\}$ of real-valued functions, all with the same domain D, *converges pointwise* to $f: D \to \mathfrak{R}$ if $\lim_{n \to \infty} f_n(x) = f(x)$ for all $x \in D$; i.e.

$$(\forall x \in D)(\forall \epsilon > 0)(\exists N)(\forall n > N)(|f(x) - f_n(x)| < \epsilon).$$

The sequence $\{f_n\}$ *converges uniformly* to $f: D \to \mathcal{R}$ if the N can be chosen so that it will work for all $x \in D$ at the same time. That is,

$$(\forall \epsilon > 0)(\exists N)(\forall n > N)(\forall x \in D)(|f(x) - f_n(x)| < \epsilon).$$

The sequence $\{f_n\}$ is said to be *pointwise bounded* if

$$(\forall x \in D)(\exists M)(\forall n)(|f_n(x)| < M).$$

The sequence is *uniformly bounded* if M can be chosen to work for all $x \in D$ at once:

$$(\exists M)(\forall n)(\forall x \in D)(|f_n(x)| < M).$$

Bibliography

Apostol, T., *Mathematical Analysis*, Reading, Mass., Addison-Wesley, 1957.

Asplund, E., L. Bungart, *A First Course in Integration*, Holt, Rinehart, and Winston, New York, 1966.

Bartle, R., *The Elements of Integration*, John Wiley and Sons, New York, 1966.

Burkill, J., *The Lebesgue Integral*, Cambridge Tracts No. 40, Cambridge University Press, New York, 1961.

Goldberg, R., *Methods of Real Analysis*, Waltham, Mass., Blaisdell, 1964.

Halmos, P., *Measure Theory*, Princeton, D. Van Nostrand, 1950.

Hewitt, E., K. Stromberg, *Real and Abstract Analysis*, New York, Springer-Verlag, 1965.

Lebesgue, H., *Measure and the Integral*, San Francisco, Holden-Day, 1966.

Munroe, M., *Introduction to Measure and Integration*, Reading, Mass., Addison-Wesley, 1953.

Royden, H., *Real Analysis (2nd Edition)*, New York, Macmillan, 1968.

Rudin, W., *Principles of Mathematical Analysis*, New York, McGraw-Hill, 1953.

Scanlon, J., *Advanced Calculus*, Boston, Heath, 1967.

Sprecher, D., *Elements of Real Analysis*, New York, Academic Press, 1970.

Temple, G., *The Structure of Lebesgue Integration Theory*, Oxford, Clarendon Press, 1971.

Williamson, J., *Lebesgue Integration*, New York, Holt, Rinehart and Winston, 1962.

INDEX

A CATALOG OF SELECTED

DOVER BOOKS
IN SCIENCE AND MATHEMATICS

A CATALOG OF SELECTED

DOVER BOOKS

IN SCIENCE AND MATHEMATICS

QUALITATIVE THEORY OF DIFFERENTIAL EQUATIONS, V.V. Nemytskii and V.V. Stepanov. Classic graduate-level text by two prominent Soviet mathematicians covers classical differential equations as well as topological dynamics and ergodic theory. Bibliographies. 523pp. 5⅜ x 8½. 65954-2 Pa. $14.95

MATRICES AND LINEAR ALGEBRA, Hans Schneider and George Phillip Barker. Basic textbook covers theory of matrices and its applications to systems of linear equations and related topics such as determinants, eigenvalues and differential equations. Numerous exercises. 432pp. 5⅜ x 8½. 66014-1 Pa. $10.95

QUANTUM THEORY, David Bohm. This advanced undergraduate-level text presents the quantum theory in terms of qualitative and imaginative concepts, followed by specific applications worked out in mathematical detail. Preface. Index. 655pp. 5⅜ x 8½. 65969-0 Pa. $14.95

ATOMIC PHYSICS (8th edition), Max Born. Nobel laureate's lucid treatment of kinetic theory of gases, elementary particles, nuclear atom, wave-corpuscles, atomic structure and spectral lines, much more. Over 40 appendices, bibliography. 495pp. 5⅜ x 8½. 65984-4 Pa. $13.95

ELECTRONIC STRUCTURE AND THE PROPERTIES OF SOLIDS: The Physics of the Chemical Bond, Walter A. Harrison. Innovative text offers basic understanding of the electronic structure of covalent and ionic solids, simple metals, transition metals and their compounds. Problems. 1980 edition. 582pp. 6⅛ x 9¼. 66021-4 Pa. $16.95

BOUNDARY VALUE PROBLEMS OF HEAT CONDUCTION, M. Necati Özisik. Systematic, comprehensive treatment of modern mathematical methods of solving problems in heat conduction and diffusion. Numerous examples and problems. Selected references. Appendices. 505pp. 5⅜ x 8½. 65990-9 Pa. $12.95

A SHORT HISTORY OF CHEMISTRY (3rd edition), J.R. Partington. Classic exposition explores origins of chemistry, alchemy, early medical chemistry, nature of atmosphere, theory of valency, laws and structure of atomic theory, much more. 428pp. 5⅜ x 8½. (Available in U.S. only) 65977-1 Pa. $11.95

A HISTORY OF ASTRONOMY, A. Pannekoek. Well-balanced, carefully reasoned study covers such topics as Ptolemaic theory, work of Copernicus, Kepler, Newton, Eddington's work on stars, much more. Illustrated. References. 521pp. 5⅜ x 8½. 65994-1 Pa. $12.95

PRINCIPLES OF METEOROLOGICAL ANALYSIS, Walter J. Saucier. Highly respected, abundantly illustrated classic reviews atmospheric variables, hydrostatics, static stability, various analyses (scalar, cross-section, isobaric, isentropic, more). For intermediate meteorology students. 454pp. 6⅛ x 9¼. 65979-8 Pa. $14.95

RELATIVITY, THERMODYNAMICS AND COSMOLOGY, Richard C. Tolman. Landmark study extends thermodynamics to special, general relativity; also applications of relativistic mechanics, thermodynamics to cosmological models. 501pp. 5⅜ x 8½. 65383-8 Pa. $13.95

APPLIED ANALYSIS, Cornelius Lanczos. Classic work on analysis and design of finite processes for approximating solution of analytical problems. Algebraic equations, matrices, harmonic analysis, quadrature methods, much more. 559pp. 5⅜ x 8½. 65656-X Pa. $13.95

INTRODUCTION TO ANALYSIS, Maxwell Rosenlicht. Unusually clear, accessible coverage of set theory, real number system, metric spaces, continuous functions, Riemann integration, multiple integrals, more. Wide range of problems. Undergraduate level. Bibliography. 254pp. 5⅜ x 8½. 65038-3 Pa. $8.95

INTRODUCTION TO QUANTUM MECHANICS With Applications to Chemistry, Linus Pauling & E. Bright Wilson, Jr. Classic undergraduate text by Nobel Prize winner applies quantum mechanics to chemical and physical problems. Numerous tables and figures enhance the text. Chapter bibliographies. Appendices. Index. 468pp. 5⅜ x 8½. 64871-0 Pa. $12.95

ASYMPTOTIC EXPANSIONS OF INTEGRALS, Norman Bleistein & Richard A. Handelsman. Best introduction to important field with applications in a variety of scientific disciplines. New preface. Problems. Diagrams. Tables. Bibliography. Index. 448pp. 5⅜ x 8½. 65082-0 Pa. $12.95

MATHEMATICS APPLIED TO CONTINUUM MECHANICS, Lee A. Segel. Analyzes models of fluid flow and solid deformation. For upper-level math, science and engineering students. 608pp. 5⅜ x 8½. 65369-2 Pa. $14.95

ELEMENTS OF REAL ANALYSIS, David A. Sprecher. Classic text covers fundamental concepts, real number system, point sets, functions of a real variable, Fourier series, much more. Over 500 exercises. 352pp. 5⅜ x 8½. 65385-4 Pa. $11.95

PHYSICAL PRINCIPLES OF THE QUANTUM THEORY, Werner Heisenberg. Nobel Laureate discusses quantum theory, uncertainty, wave mechanics, work of Dirac, Schroedinger, Compton, Wilson, Einstein, etc. 184pp. 5⅜ x 8½. 60113-7 Pa. $6.95

INTRODUCTORY REAL ANALYSIS, A.N. Kolmogorov, S.V. Fomin. Translated by Richard A. Silverman. Self-contained, evenly paced introduction to real and functional analysis. Some 350 problems. 403pp. 5⅜ x 8½. 61226-0 Pa. $10.95

PROBLEMS AND SOLUTIONS IN QUANTUM CHEMISTRY AND PHYSICS, Charles S. Johnson, Jr. and Lee G. Pedersen. Unusually varied problems, detailed solutions in coverage of quantum mechanics, wave mechanics, angular momentum, molecular spectroscopy, scattering theory, more. 280 problems plus 139 supplementary exercises. 430pp. 6½ x 9¼. 65236-X Pa. $13.95

ASYMPTOTIC METHODS IN ANALYSIS, N.G. de Bruijn. An inexpensive, comprehensive guide to asymptotic methods—the pioneering work that teaches by explaining worked examples in detail. Index. 224pp. 5⅜ x 8½.   64221-6 Pa. $7.95

OPTICAL RESONANCE AND TWO-LEVEL ATOMS, L. Allen and J. H. Eberly. Clear, comprehensive introduction to basic principles behind all quantum optical resonance phenomena. 53 illustrations. Preface. Index. 256pp. 5⅜ x 8½.
65533-4 Pa. $8.95

COMPLEX VARIABLES, Francis J. Flanigan. Unusual approach, delaying complex algebra till harmonic functions have been analyzed from real variable viewpoint. Includes problems with answers. 364pp. 5⅜ x 8½.   61388-7 Pa. $9.95

ATOMIC SPECTRA AND ATOMIC STRUCTURE, Gerhard Herzberg. One of best introductions; especially for specialist in other fields. Treatment is physical rather than mathematical. 80 illustrations. 257pp. 5⅜ x 8½.   60115-3 Pa. $7.95

APPLIED COMPLEX VARIABLES, John W. Dettman. Step-by-step coverage of fundamentals of analytic function theory—plus lucid exposition of five important applications: Potential Theory; Ordinary Differential Equations; Fourier Transforms; Laplace Transforms; Asymptotic Expansions. 66 figures. Exercises at chapter ends. 512pp. 5⅜ x 8½.   64670-X Pa. $12.95

ULTRASONIC ABSORPTION: An Introduction to the Theory of Sound Absorption and Dispersion in Gases, Liquids and Solids, A.B. Bhatia. Standard reference in the field provides a clear, systematically organized introductory review of fundamental concepts for advanced graduate students, research workers. Numerous diagrams. Bibliography. 440pp. 5⅜ x 8½.   64917-2 Pa. $11.95

UNBOUNDED LINEAR OPERATORS: Theory and Applications, Seymour Goldberg. Classic presents systematic treatment of the theory of unbounded linear operators in normed linear spaces with applications to differential equations. Bibliography. 199pp. 5⅜ x 8½.   64830-3 Pa. $7.95

LIGHT SCATTERING BY SMALL PARTICLES, H.C. van de Hulst. Comprehensive treatment including full range of useful approximation methods for researchers in chemistry, meteorology and astronomy. 44 illustrations. 470pp. 5⅜ x 8½.
64228-3 Pa. $12.95

CONFORMAL MAPPING ON RIEMANN SURFACES, Harvey Cohn. Lucid, insightful book presents ideal coverage of subject. 334 exercises make book perfect for self-study. 55 figures. 352pp. 5⅞ x 8¼.   64025-6 Pa. $11.95

OPTICKS, Sir Isaac Newton. Newton's own experiments with spectroscopy, colors, lenses, reflection, refraction, etc., in language the layman can follow. Foreword by Albert Einstein. 532pp. 5⅜ x 8½.   60205-2 Pa. $12.95

GENERALIZED INTEGRAL TRANSFORMATIONS, A.H. Zemanian. Graduate-level study of recent generalizations of the Laplace, Mellin, Hankel, K. Weierstrass, convolution and other simple transformations. Bibliography. 320pp. 5⅜ x 8½.
65375-7 Pa. $8.95

THE ELECTROMAGNETIC FIELD, Albert Shadowitz. Comprehensive undergraduate text covers basics of electric and magnetic fields, builds up to electromagnetic theory. Also related topics, including relativity. Over 900 problems. 768pp. 5⅝ x 8¼. 65660-8 Pa. $18.95

FOURIER SERIES, Georgi P. Tolstov. Translated by Richard A. Silverman. A valuable addition to the literature on the subject, moving clearly from subject to subject and theorem to theorem. 107 problems, answers. 336pp. 5⅜ x 8½. 63317-9 Pa. $9.95

THEORY OF ELECTROMAGNETIC WAVE PROPAGATION, Charles Herach Papas. Graduate-level study discusses the Maxwell field equations, radiation from wire antennas, the Doppler effect and more. xiii + 244pp. 5⅜ x 8½. 65678-0 Pa. $6.95

DISTRIBUTION THEORY AND TRANSFORM ANALYSIS: An Introduction to Generalized Functions, with Applications, A.H. Zemanian. Provides basics of distribution theory, describes generalized Fourier and Laplace transformations. Numerous problems. 384pp. 5⅜ x 8½. 65479-6 Pa. $11.95

THE PHYSICS OF WAVES, William C. Elmore and Mark A. Heald. Unique overview of classical wave theory. Acoustics, optics, electromagnetic radiation, more. Ideal as classroom text or for self-study. Problems. 477pp. 5⅜ x 8½. 64926-1 Pa. $13.95

CALCULUS OF VARIATIONS WITH APPLICATIONS, George M. Ewing. Applications-oriented introduction to variational theory develops insight and promotes understanding of specialized books, research papers. Suitable for advanced undergraduate/graduate students as primary, supplementary text. 352pp. 5⅜ x 8½. 64856-7 Pa. $9.95

A TREATISE ON ELECTRICITY AND MAGNETISM, James Clerk Maxwell. Important foundation work of modern physics. Brings to final form Maxwell's theory of electromagnetism and rigorously derives his general equations of field theory. 1,084pp. 5⅜ x 8½. 60636-8, 60637-6 Pa., Two-vol. set $25.90

AN INTRODUCTION TO THE CALCULUS OF VARIATIONS, Charles Fox. Graduate-level text covers variations of an integral, isoperimetrical problems, least action, special relativity, approximations, more. References. 279pp. 5⅜ x 8½. 65499-0 Pa. $8.95

HYDRODYNAMIC AND HYDROMAGNETIC STABILITY, S. Chandrasekhar. Lucid examination of the Rayleigh-Benard problem; clear coverage of the theory of instabilities causing convection. 704pp. 5⅜ x 8¼. 64071-X Pa. $14.95

CALCULUS OF VARIATIONS, Robert Weinstock. Basic introduction covering isoperimetric problems, theory of elasticity, quantum mechanics, electrostatics, etc. Exercises throughout. 326pp. 5⅜ x 8½. 63069-2 Pa. $9.95

DYNAMICS OF FLUIDS IN POROUS MEDIA, Jacob Bear. For advanced students of ground water hydrology, soil mechanics and physics, drainage and irrigation engineering and more. 335 illustrations. Exercises, with answers. 784pp. 6⅛ x 9¼. 65675-6 Pa. $19.95

NUMERICAL METHODS FOR SCIENTISTS AND ENGINEERS, Richard Hamming. Classic text stresses frequency approach in coverage of algorithms, polynomial approximation, Fourier approximation, exponential approximation, other topics. Revised and enlarged 2nd edition. 721pp. 5⅜ x 8½.   65241-6 Pa. $15.95

THEORETICAL SOLID STATE PHYSICS, Vol. 1: Perfect Lattices in Equilibrium; Vol. II: Non-Equilibrium and Disorder, William Jones and Norman H. March. Monumental reference work covers fundamental theory of equilibrium properties of perfect crystalline solids, non-equilibrium properties, defects and disordered systems. Appendices. Problems. Preface. Diagrams. Index. Bibliography. Total of 1,301pp. 5⅜ x 8½. Two volumes.   Vol. I: 65015-4 Pa. $16.95
Vol. II: 65016-2 Pa. $16.95

OPTIMIZATION THEORY WITH APPLICATIONS, Donald A. Pierre. Broad spectrum approach to important topic. Classical theory of minima and maxima, calculus of variations, simplex technique and linear programming, more. Many problems, examples. 640pp. 5⅜ x 8½.   65205-X Pa. $16.95

THE CONTINUUM: A Critical Examination of the Foundation of Analysis, Hermann Weyl. Classic of 20th-century foundational research deals with the conceptual problem posed by the continuum. 156pp. 5⅜ x 8½.   67982-9 Pa. $6.95

ESSAYS ON THE THEORY OF NUMBERS, Richard Dedekind. Two classic essays by great German mathematician: on the theory of irrational numbers; and on transfinite numbers and properties of natural numbers. 115pp. 5⅜ x 8½.
21010-3 Pa. $5.95

THE FUNCTIONS OF MATHEMATICAL PHYSICS, Harry Hochstadt. Comprehensive treatment of orthogonal polynomials, hypergeometric functions, Hill's equation, much more. Bibliography. Index. 322pp. 5⅜ x 8½. 65214-9 Pa. $9.95

NUMBER THEORY AND ITS HISTORY, Oystein Ore. Unusually clear, accessible introduction covers counting, properties of numbers, prime numbers, much more. Bibliography. 380pp. 5⅜ x 8½.   65620-9 Pa. $10.95

THE VARIATIONAL PRINCIPLES OF MECHANICS, Cornelius Lanczos. Graduate level coverage of calculus of variations, equations of motion, relativistic mechanics, more. First inexpensive paperbound edition of classic treatise. Index. Bibliography. 418pp. 5⅜ x 8½.   65067-7 Pa. $12.95

MATHEMATICAL TABLES AND FORMULAS, Robert D. Carmichael and Edwin R. Smith. Logarithms, sines, tangents, trig functions, powers, roots, reciprocals, exponential and hyperbolic functions, formulas and theorems. 269pp. 5⅜ x 8½.
60111-0 Pa. $6.95

THEORETICAL PHYSICS, Georg Joos, with Ira M. Freeman. Classic overview covers essential math, mechanics, electromagnetic theory, thermodynamics, quantum mechanics, nuclear physics, other topics. First paperback edition. xxiii + 885pp. 5⅜ x 8½.   65227-0 Pa. $21.95

HANDBOOK OF MATHEMATICAL FUNCTIONS WITH FORMULAS, GRAPHS, AND MATHEMATICAL TABLES, edited by Milton Abramowitz and Irene A. Stegun. Vast compendium: 29 sets of tables, some to as high as 20 places. 1,046pp. 8 x 10½. 61272-4 Pa. $26.95

MATHEMATICAL METHODS IN PHYSICS AND ENGINEERING, John W. Dettman. Algebraically based approach to vectors, mapping, diffraction, other topics in applied math. Also generalized functions, analytic function theory, more. Exercises. 448pp. 5⅜ x 8¼. 65649-7 Pa. $10.95

A SURVEY OF NUMERICAL MATHEMATICS, David M. Young and Robert Todd Gregory. Broad self-contained coverage of computer-oriented numerical algorithms for solving various types of mathematical problems in linear algebra, ordinary and partial, differential equations, much more. Exercises. Total of 1,248pp. 5⅜ x 8½. Two volumes. Vol. I: 65691-8 Pa. $16.95
Vol. II: 65692-6 Pa. $16.95

TENSOR ANALYSIS FOR PHYSICISTS, J.A. Schouten. Concise exposition of the mathematical basis of tensor analysis, integrated with well-chosen physical examples of the theory. Exercises. Index. Bibliography. 289pp. 5⅜ x 8½. 65582-2 Pa. $8.95

INTRODUCTION TO NUMERICAL ANALYSIS (2nd Edition), F.B. Hildebrand. Classic, fundamental treatment covers computation, approximation, interpolation, numerical differentiation and integration, other topics. 150 new problems. 669pp. 5⅜ x 8½. 65363-3 Pa. $16.95

INVESTIGATIONS ON THE THEORY OF THE BROWNIAN MOVEMENT, Albert Einstein. Five papers (1905–8) investigating dynamics of Brownian motion and evolving elementary theory. Notes by R. Fürth. 122pp. 5⅜ x 8½.
60304-0 Pa. $5.95

CATASTROPHE THEORY FOR SCIENTISTS AND ENGINEERS, Robert Gilmore. Advanced-level treatment describes mathematics of theory grounded in the work of Poincaré, R. Thom, other mathematicians. Also important applications to problems in mathematics, physics, chemistry and engineering. 1981 edition. References. 28 tables. 397 black-and-white illustrations. xvii + 666pp. 6⅛ x 9¼.
67539-4 Pa. $17.95

AN INTRODUCTION TO STATISTICAL THERMODYNAMICS, Terrell L. Hill. Excellent basic text offers wide-ranging coverage of quantum statistical mechanics, systems of interacting molecules, quantum statistics, more. 523pp. 5⅜ x 8½.
65242-4 Pa. $12.95

STATISTICAL PHYSICS, Gregory H. Wannier. Classic text combines thermodynamics, statistical mechanics and kinetic theory in one unified presentation of thermal physics. Problems with solutions. Bibliography. 532pp. 5⅜ x 8½.
65401-X Pa. $12.95

ORDINARY DIFFERENTIAL EQUATIONS, Morris Tenenbaum and Harry Pollard. Exhaustive survey of ordinary differential equations for undergraduates in mathematics, engineering, science. Thorough analysis of theorems. Diagrams. Bibliography. Index. 818pp. 5⅜ x 8½. 64940-7 Pa. $18.95

STATISTICAL MECHANICS: Principles and Applications, Terrell L. Hill. Standard text covers fundamentals of statistical mechanics, applications to fluctuation theory, imperfect gases, distribution functions, more. 448pp. 5⅜ x 8½. 65390-0 Pa. $11.95

ORDINARY DIFFERENTIAL EQUATIONS AND STABILITY THEORY: An Introduction, David A. Sánchez. Brief, modern treatment. Linear equation, stability theory for autonomous and nonautonomous systems, etc. 164pp. 5⅜ x 8¼. 63828-6 Pa. $6.95

THIRTY YEARS THAT SHOOK PHYSICS: The Story of Quantum Theory, George Gamow. Lucid, accessible introduction to influential theory of energy and matter. Careful explanations of Dirac's anti-particles, Bohr's model of the atom, much more. 12 plates. Numerous drawings. 240pp. 5⅜ x 8½. 24895-X Pa. $7.95

THEORY OF MATRICES, Sam Perlis. Outstanding text covering rank, nonsingularity and inverses in connection with the development of canonical matrices under the relation of equivalence, and without the intervention of determinants. Includes exercises. 237pp. 5⅜ x 8½. 66810-X Pa. $8.95

GREAT EXPERIMENTS IN PHYSICS: Firsthand Accounts from Galileo to Einstein, edited by Morris H. Shamos. 25 crucial discoveries: Newton's laws of motion, Chadwick's study of the neutron, Hertz on electromagnetic waves, more. Original accounts clearly annotated. 370pp. 5⅜ x 8½. 25346-5 Pa. $10.95

INTRODUCTION TO PARTIAL DIFFERENTIAL EQUATIONS WITH APPLICATIONS, E.C. Zachmanoglou and Dale W. Thoe. Essentials of partial differential equations applied to common problems in engineering and the physical sciences. Problems and answers. 416pp. 5⅜ x 8½. 65251-3 Pa. $11.95

BURNHAM'S CELESTIAL HANDBOOK, Robert Burnham, Jr. Thorough guide to the stars beyond our solar system. Exhaustive treatment. Alphabetical by constellation: Andromeda to Cetus in Vol. 1; Chamaeleon to Orion in Vol. 2; and Pavo to Vulpecula in Vol. 3. Hundreds of illustrations. Index in Vol. 3. 2,000pp. 6⅛ x 9¼. 23567-X, 23568-8, 23673-0 Pa., Three-vol. set $44.85

CHEMICAL MAGIC, Leonard A. Ford. Second Edition, Revised by E. Winston Grundmeier. Over 100 unusual stunts demonstrating cold fire, dust explosions, much more. Text explains scientific principles and stresses safety precautions. 128pp. 5⅜ x 8½. 67628-5 Pa. $5.95

AMATEUR ASTRONOMER'S HANDBOOK, J.B. Sidgwick. Timeless, comprehensive coverage of telescopes, mirrors, lenses, mountings, telescope drives, micrometers, spectroscopes, more. 189 illustrations. 576pp. 5⅜ x 8¼. (Available in U.S. only) 24034-7 Pa. $11.95

SPECIAL FUNCTIONS, N.N. Lebedev. Translated by Richard Silverman. Famous Russian work treating more important special functions, with applications to specific problems of physics and engineering. 38 figures. 308pp. 5⅜ x 8½. 60624-4 Pa. $9.95

OBSERVATIONAL ASTRONOMY FOR AMATEURS, J.B. Sidgwick. Mine of useful data for observation of sun, moon, planets, asteroids, aurorae, meteors, comets, variables, binaries, etc. 39 illustrations. 384pp. 5⅜ x 8¼. (Available in U.S. only) 24033-9 Pa. $8.95

INTEGRAL EQUATIONS, F.G. Tricomi. Authoritative, well-written treatment of extremely useful mathematical tool with wide applications. Volterra Equations, Fredholm Equations, much more. Advanced undergraduate to graduate level. Exercises. Bibliography. 238pp. 5⅜ x 8½. 64828-1 Pa. $8.95

POPULAR LECTURES ON MATHEMATICAL LOGIC, Hao Wang. Noted logician's lucid treatment of historical developments, set theory, model theory, recursion theory and constructivism, proof theory, more. 3 appendixes. Bibliography. 1981 edition. ix + 283pp. 5⅜ x 8½. 67632-3 Pa. $8.95

MODERN NONLINEAR EQUATIONS, Thomas L. Saaty. Emphasizes practical solution of problems; covers seven types of equations. ". . . a welcome contribution to the existing literature...."–*Math Reviews.* 490pp. 5⅜ x 8½. 64232-1 Pa. $13.95

FUNDAMENTALS OF ASTRODYNAMICS, Roger Bate et al. Modern approach developed by U.S. Air Force Academy. Designed as a first course. Problems, exercises. Numerous illustrations. 455pp. 5⅜ x 8½. 60061-0 Pa. $10.95

INTRODUCTION TO LINEAR ALGEBRA AND DIFFERENTIAL EQUATIONS, John W. Dettman. Excellent text covers complex numbers, determinants, orthonormal bases, Laplace transforms, much more. Exercises with solutions. Undergraduate level. 416pp. 5⅜ x 8½. 65191-6 Pa. $11.95

INCOMPRESSIBLE AERODYNAMICS, edited by Bryan Thwaites. Covers theoretical and experimental treatment of the uniform flow of air and viscous fluids past two-dimensional aerofoils and three-dimensional wings; many other topics. 654pp. 5⅜ x 8½. 65465-6 Pa. $16.95

INTRODUCTION TO DIFFERENCE EQUATIONS, Samuel Goldberg. Exceptionally clear exposition of important discipline with applications to sociology, psychology, economics. Many illustrative examples; over 250 problems. 260pp. 5⅜ x 8½. 65084-7 Pa. $8.95

LAMINAR BOUNDARY LAYERS, edited by L. Rosenhead. Engineering classic covers steady boundary layers in two- and three- dimensional flow, unsteady boundary layers, stability, observational techniques, much more. 708pp. 5⅜ x 8½. 65646-2 Pa. $18 95

LECTURES ON CLASSICAL DIFFERENTIAL GEOMETRY, Second Edition, Dirk J. Struik. Excellent brief introduction covers curves, theory of surfaces, fundamental equations, geometry on a surface, conformal mapping, other topics. Problems. 240pp. 5⅜ x 8½. 65609-8 Pa. $8.95

ROTARY-WING AERODYNAMICS, W.Z. Stepniewski. Clear, concise text covers aerodynamic phenomena of the rotor and offers guidelines for helicopter performance evaluation. Originally prepared for NASA. 537 figures. 640pp. 6⅛ x 9¼.
64647-5 Pa. $16.95

DIFFERENTIAL GEOMETRY, Heinrich W. Guggenheimer. Local differential geometry as an application of advanced calculus and linear algebra. Curvature, transformation groups, surfaces, more. Exercises. 62 figures. 378pp. 5⅜ x 8½.
63433-7 Pa. $9.95

INTRODUCTION TO SPACE DYNAMICS, William Tyrrell Thomson. Comprehensive, classic introduction to space-flight engineering for advanced undergraduate and graduate students. Includes vector algebra, kinematics, transformation of coordinates. Bibliography. Index. 352pp. 5⅜ x 8½.
65113-4 Pa. $9.95

A SURVEY OF MINIMAL SURFACES, Robert Osserman. Up-to-date, in-depth discussion of the field for advanced students. Corrected and enlarged edition covers new developments. Includes numerous problems. 192pp. 5⅜ x 8½. 64998-9 Pa. $8.95

ANALYTICAL MECHANICS OF GEARS, Earle Buckingham. Indispensable reference for modern gear manufacture covers conjugate gear-tooth action, gear-tooth profiles of various gears, many other topics. 263 figures. 102 tables. 546pp. 5⅜ x 8½.
65712-4 Pa. $14.95

SET THEORY AND LOGIC, Robert R. Stoll. Lucid introduction to unified theory of mathematical concepts. Set theory and logic seen as tools for conceptual understanding of real number system. 496pp. 5⅜ x 8¼. 63829-4 Pa. $12.95

A HISTORY OF MECHANICS, René Dugas. Monumental study of mechanical principles from antiquity to quantum mechanics. Contributions of ancient Greeks, Galileo, Leonardo, Kepler, Lagrange, many others. 671pp. 5⅜ x 8½.
65632-2 Pa. $14.95

FAMOUS PROBLEMS OF GEOMETRY AND HOW TO SOLVE THEM, Benjamin Bold. Squaring the circle, trisecting the angle, duplicating the cube: learn their history, why they are impossible to solve, then solve them yourself. 128pp. 5⅜ x 8½. 24297-8 Pa. $4.95

MECHANICAL VIBRATIONS, J.P. Den Hartog. Classic textbook offers lucid explanations and illustrative models, applying theories of vibrations to a variety of practical industrial engineering problems. Numerous figures. 233 problems, solutions. Appendix. Index. Preface. 436pp. 5⅜ x 8½. 64785-4 Pa. $11.95

CURVATURE AND HOMOLOGY, Samuel I. Goldberg. Thorough treatment of specialized branch of differential geometry. Covers Riemannian manifolds, topology of differentiable manifolds, compact Lie groups, other topics. Exercises. 315pp. 5⅜ x 8½. 64314-X Pa. $9.95

HISTORY OF STRENGTH OF MATERIALS, Stephen P. Timoshenko. Excellent historical survey of the strength of materials with many references to the theories of elasticity and structure. 245 figures. 452pp. 5⅜ x 8½. 61187-6 Pa. $12.95

GEOMETRY OF COMPLEX NUMBERS, Hans Schwerdtfeger. Illuminating, widely praised book on analytic geometry of circles, the Moebius transformation, and two-dimensional non-Euclidean geometries. 200pp. 5⅝ x 8¼. 63830-8 Pa. $8.95

MECHANICS, J.P. Den Hartog. A classic introductory text or refresher. Hundreds of applications and design problems illuminate fundamentals of trusses, loaded beams and cables, etc. 334 answered problems. 462pp. 5⅜ x 8½. 60754-2 Pa. $11.95

TOPOLOGY, John G. Hocking and Gail S. Young. Superb one-year course in classical topology. Topological spaces and functions, point-set topology, much more. Examples and problems. Bibliography. Index. 384pp. 5⅝ x 8¼. 65676-4 Pa. $10.95

STRENGTH OF MATERIALS, J.P. Den Hartog. Full, clear treatment of basic material (tension, torsion, bending, etc.) plus advanced material on engineering methods, applications. 350 answered problems. 323pp. 5⅜ x 8½. 60755-0 Pa. $9.95

ELEMENTARY CONCEPTS OF TOPOLOGY, Paul Alexandroff. Elegant, intuitive approach to topology from set-theoretic topology to Betti groups; how concepts of topology are useful in math and physics. 25 figures. 57pp. 5⅜ x 8½. 60747-X Pa. $3.95

ADVANCED STRENGTH OF MATERIALS, J.P. Den Hartog. Superbly written advanced text covers torsion, rotating disks, membrane stresses in shells, much more. Many problems and answers. 388pp. 5⅜ x 8½. 65407-9 Pa. $10.95

COMPUTABILITY AND UNSOLVABILITY, Martin Davis. Classic graduate-level introduction to theory of computability, usually referred to as theory of recurrent functions. New preface and appendix. 288pp. 5⅜ x 8½. 61471-9 Pa. $8.95

GENERAL CHEMISTRY, Linus Pauling. Revised 3rd edition of classic first-year text by Nobel laureate. Atomic and molecular structure, quantum mechanics, statistical mechanics, thermodynamics correlated with descriptive chemistry. Problems. 992pp. 5⅜ x 8½. 65622-5 Pa. $19.95

AN INTRODUCTION TO MATRICES, SETS AND GROUPS FOR SCIENCE STUDENTS, G. Stephenson. Concise, readable text introduces sets, groups, and most importantly, matrices to undergraduate students of physics, chemistry, and engineering. Problems. 164pp. 5⅜ x 8½. 65077-4 Pa. $7.95

THE HISTORICAL BACKGROUND OF CHEMISTRY, Henry M. Leicester. Evolution of ideas, not individual biography. Concentrates on formulation of a coherent set of chemical laws. 260pp. 5⅜ x 8½. 61053-5 Pa. $8.95

THE PHILOSOPHY OF MATHEMATICS: An Introductory Essay, Stephan Körner. Surveys the views of Plato, Aristotle, Leibniz & Kant concerning propositions and theories of applied and pure mathematics. Introduction. Two appendices. Index. 198pp. 5⅜ x 8½. 25048-2 Pa. $8.95

THE DEVELOPMENT OF MODERN CHEMISTRY, Aaron J. Ihde. Authoritative history of chemistry from ancient Greek theory to 20th-century innovation. Covers major chemists and their discoveries. 209 illustrations. 14 tables. Bibliographies. Indices. Appendices. 851pp. 5⅜ x 8½. 64235-6 Pa. $18.95

DE RE METALLICA, Georgius Agricola. The famous Hoover translation of greatest treatise on technological chemistry, engineering, geology, mining of early modern times (1556). All 289 original woodcuts. 638pp. 6¾ x 11. 60006-8 Pa. $21.95

SOME THEORY OF SAMPLING, William Edwards Deming. Analysis of the problems, theory and design of sampling techniques for social scientists, industrial managers and others who find statistics increasingly important in their work. 61 tables. 90 figures. xvii + 602pp. 5⅜ x 8½. 64684-X Pa. $16.95

THE VARIOUS AND INGENIOUS MACHINES OF AGOSTINO RAMELLI: A Classic Sixteenth-Century Illustrated Treatise on Technology, Agostino Ramelli. One of the most widely known and copied works on machinery in the 16th century. 194 detailed plates of water pumps, grain mills, cranes, more. 608pp. 9 x 12. 28180-9 Pa. $24.95

LINEAR PROGRAMMING AND ECONOMIC ANALYSIS, Robert Dorfman, Paul A. Samuelson and Robert M. Solow. First comprehensive treatment of linear programming in standard economic analysis. Game theory, modern welfare economics, Leontief input-output, more. 525pp. 5⅜ x 8½. 65491-5 Pa. $14.95

ELEMENTARY DECISION THEORY, Herman Chernoff and Lincoln E. Moses. Clear introduction to statistics and statistical theory covers data processing, probability and random variables, testing hypotheses, much more. Exercises. 364pp. 5⅜ x 8½. 65218-1 Pa. $10.95

THE COMPLEAT STRATEGYST: Being a Primer on the Theory of Games of Strategy, J.D. Williams. Highly entertaining classic describes, with many illustrated examples, how to select best strategies in conflict situations. Prefaces. Appendices. 268pp. 5⅜ x 8½. 25101-2 Pa. $7.95

CONSTRUCTIONS AND COMBINATORIAL PROBLEMS IN DESIGN OF EXPERIMENTS, Damaraju Raghavarao. In-depth reference work examines orthogonal Latin squares, incomplete block designs, tactical configuration, partial geometry, much more. Abundant explanations, examples. 416pp. 5⅜ x 8¼. 65685-3 Pa. $10.95

THE ABSOLUTE DIFFERENTIAL CALCULUS (CALCULUS OF TENSORS), Tullio Levi-Civita. Great 20th-century mathematician's classic work on material necessary for mathematical grasp of theory of relativity. 452pp. 5⅜ x 8½. 63401-9 Pa. $11.95

VECTOR AND TENSOR ANALYSIS WITH APPLICATIONS, A.I. Borisenko and I.E. Tarapov. Concise introduction. Worked-out problems, solutions, exercises. 257pp. 5⅜ x 8¼. 63833-2 Pa. $8.95

THE FOUR-COLOR PROBLEM: Assaults and Conquest, Thomas L. Saaty and Paul G. Kainen. Engrossing, comprehensive account of the century-old combinatorial topological problem, its history and solution. Bibliographies. Index. 110 figures. 228pp. 5⅜ x 8½. 65092-8 Pa. $7.95

CATALYSIS IN CHEMISTRY AND ENZYMOLOGY, William P. Jencks. Exceptionally clear coverage of mechanisms for catalysis, forces in aqueous solution, carbonyl- and acyl-group reactions, practical kinetics, more. 864pp. 5⅜ x 8½.
65460-5 Pa. $19.95

PROBABILITY: An Introduction, Samuel Goldberg. Excellent basic text covers set theory, probability theory for finite sample spaces, binomial theorem, much more. 360 problems. Bibliographies. 322pp. 5⅜ x 8½.
65252-1 Pa. $10.95

LIGHTNING, Martin A. Uman. Revised, updated edition of classic work on the physics of lightning. Phenomena, terminology, measurement, photography, spectroscopy, thunder, more. Reviews recent research. Bibliography. Indices. 320pp. 5⅜ x 8¼.
64575-4 Pa. $8.95

PROBABILITY THEORY: A Concise Course, Y.A. Rozanov. Highly readable, self-contained introduction covers combination of events, dependent events, Bernoulli trials, etc. Translation by Richard Silverman. 148pp. 5⅜ x 8¼.
63544-9 Pa. $7.95

AN INTRODUCTION TO HAMILTONIAN OPTICS, H. A. Buchdahl. Detailed account of the Hamiltonian treatment of aberration theory in geometrical optics. Many classes of optical systems defined in terms of the symmetries they possess. Problems with detailed solutions. 1970 edition. xv + 360pp. 5⅜ x 8½.
67597-1 Pa. $10.95

STATISTICS MANUAL, Edwin L. Crow, et al. Comprehensive, practical collection of classical and modern methods prepared by U.S. Naval Ordnance Test Station. Stress on use. Basics of statistics assumed. 288pp. 5⅜ x 8½.
60599-X Pa. $7.95

DICTIONARY/OUTLINE OF BASIC STATISTICS, John E. Freund and Frank J. Williams. A clear concise dictionary of over 1,000 statistical terms and an outline of statistical formulas covering probability, nonparametric tests, much more. 208pp. 5⅜ x 8½.
66796-0 Pa. $7.95

STATISTICAL METHOD FROM THE VIEWPOINT OF QUALITY CONTROL, Walter A. Shewhart. Important text explains regulation of variables, uses of statistical control to achieve quality control in industry, agriculture, other areas. 192pp. 5⅜ x 8½.
65232-7 Pa. $7.95

METHODS OF THERMODYNAMICS, Howard Reiss. Outstanding text focuses on physical technique of thermodynamics, typical problem areas of understanding, and significance and use of thermodynamic potential. 1965 edition. 238pp. 5⅜ x 8½.
69445-3 Pa. $8.95

STATISTICAL ADJUSTMENT OF DATA, W. Edwards Deming. Introduction to basic concepts of statistics, curve fitting, least squares solution, conditions without parameter, conditions containing parameters. 26 exercises worked out. 271pp. 5⅜ x 8½.
64685-8 Pa. $9.95

TENSOR CALCULUS, J.L. Synge and A. Schild. Widely used introductory text covers spaces and tensors, basic operations in Riemannian space, non-Riemannian spaces, etc. 324pp. 5⅜ x 8¼.
63612-7 Pa. $9.95

A CONCISE HISTORY OF MATHEMATICS, Dirk J. Struik. The best brief history of mathematics. Stresses origins and covers every major figure from ancient Near East to 19th century. 41 illustrations. 195pp. 5⅜ x 8½.   60255-9 Pa. $8.95

A SHORT ACCOUNT OF THE HISTORY OF MATHEMATICS, W.W. Rouse Ball. One of clearest, most authoritative surveys from the Egyptians and Phoenicians through 19th-century figures such as Grassman, Galois, Riemann. Fourth edition. 522pp. 5⅜ x 8½.   20630-0 Pa. $11.95

HISTORY OF MATHEMATICS, David E. Smith. Nontechnical survey from ancient Greece and Orient to late 19th century; evolution of arithmetic, geometry, trigonometry, calculating devices, algebra, the calculus. 362 illustrations. 1,355pp. 5⅜ x 8½.   20429-4, 20430-8 Pa., Two-vol. set $26.90

THE GEOMETRY OF RENÉ DESCARTES, René Descartes. The great work founded analytical geometry. Original French text, Descartes' own diagrams, together with definitive Smith-Latham translation. 244pp. 5⅜ x 8½.   60068-8 Pa. $8.95

THE ORIGINS OF THE INFINITESIMAL CALCULUS, Margaret E. Baron. Only fully detailed and documented account of crucial discipline: origins; development by Galileo, Kepler, Cavalieri; contributions of Newton, Leibniz, more. 304pp. 5⅜ x 8½. (Available in U.S. and Canada only)   65371-4 Pa. $9.95

THE HISTORY OF THE CALCULUS AND ITS CONCEPTUAL DEVELOPMENT, Carl B. Boyer. Origins in antiquity, medieval contributions, work of Newton, Leibniz, rigorous formulation. Treatment is verbal. 346pp. 5⅜ x 8½. 60509-4 Pa. $9.95

THE THIRTEEN BOOKS OF EUCLID'S ELEMENTS, translated with introduction and commentary by Sir Thomas L. Heath. Definitive edition. Textual and linguistic notes, mathematical analysis. 2,500 years of critical commentary. Not abridged. 1,414pp. 5⅜ x 8½.   60088-2, 60089-0, 60090-4 Pa., Three-vol. set $32.85

GAMES AND DECISIONS: Introduction and Critical Survey, R. Duncan Luce and Howard Raiffa. Superb nontechnical introduction to game theory, primarily applied to social sciences. Utility theory, zero-sum games, n-person games, decision-making, much more. Bibliography. 509pp. 5⅜ x 8½.   65943-7 Pa. $13.95

THE HISTORICAL ROOTS OF ELEMENTARY MATHEMATICS, Lucas N.H. Bunt, Phillip S. Jones, and Jack D. Bedient. Fundamental underpinnings of modern arithmetic, algebra, geometry and number systems derived from ancient civilizations. 320pp. 5⅜ x 8½.   25563-8 Pa. $8.95

CALCULUS REFRESHER FOR TECHNICAL PEOPLE, A. Albert Klaf. Covers important aspects of integral and differential calculus via 756 questions. 566 problems, most answered. 431pp. 5⅜ x 8½.   20370-0 Pa. $8.95

CHALLENGING MATHEMATICAL PROBLEMS WITH ELEMENTARY SOLUTIONS, A.M. Yaglom and I.M. Yaglom. Over 170 challenging problems on probability theory, combinatorial analysis, points and lines, topology, convex polygons, many other topics. Solutions. Total of 445pp. 5⅜ x 8½. Two-vol. set.

Vol. I: 65536-9 Pa. $7.95
Vol. II: 65537-7 Pa. $7.95

FIFTY CHALLENGING PROBLEMS IN PROBABILITY WITH SOLUTIONS, Frederick Mosteller. Remarkable puzzlers, graded in difficulty, illustrate elementary and advanced aspects of probability. Detailed solutions. 88pp. 5⅜ x 8½.

65355-2 Pa. $4.95

EXPERIMENTS IN TOPOLOGY, Stephen Barr. Classic, lively explanation of one of the byways of mathematics. Klein bottles, Moebius strips, projective planes, map coloring, problem of the Koenigsberg bridges, much more, described with clarity and wit. 43 figures. 210pp. 5⅜ x 8½. 25933-1 Pa. $6.95

RELATIVITY IN ILLUSTRATIONS, Jacob T. Schwartz. Clear nontechnical treatment makes relativity more accessible than ever before. Over 60 drawings illustrate concepts more clearly than text alone. Only high school geometry needed. Bibliography. 128pp. 6⅛ x 9¼. 25965-X Pa. $7.95

AN INTRODUCTION TO ORDINARY DIFFERENTIAL EQUATIONS, Earl A. Coddington. A thorough and systematic first course in elementary differential equations for undergraduates in mathematics and science, with many exercises and problems (with answers). Index. 304pp. 5⅜ x 8½. 65942-9 Pa. $8.95

FOURIER SERIES AND ORTHOGONAL FUNCTIONS, Harry F. Davis. An incisive text combining theory and practical example to introduce Fourier series, orthogonal functions and applications of the Fourier method to boundary-value problems. 570 exercises. Answers and notes. 416pp. 5⅜ x 8½. 65973-9 Pa. $11.95

AN INTRODUCTION TO ALGEBRAIC STRUCTURES, Joseph Landin. Superb self-contained text covers "abstract algebra": sets and numbers, theory of groups, theory of rings, much more. Numerous well-chosen examples, exercises. 247pp. 5⅜ x 8½.

65940-2 Pa. $8.95

STARS AND RELATIVITY, Ya. B. Zel'dovich and I. D. Novikov. Vol. 1 of *Relativistic Astrophysics* by famed Russian scientists. General relativity, properties of matter under astrophysical conditions, stars and stellar systems. Deep physical insights, clear presentation. 1971 edition. References. 544pp. 5⅜ x 8½.

69424-0 Pa. $14.95
